Synthesis Lectures on Technology Management & Entrepreneurship

In the dynamic intersection of technology innovation and entrepreneurship, organizations navigate a rapidly evolving landscape. Books in the series **Synthesis Lectures on Technology Management & Entrepreneurship** are aimed at helping readers with different backgrounds understand topics in the field. Coverage includes successful entrepreneurship, effective management of emerging technologies to deal with uncertainty and capitalize on opportunities, management of innovation processes, and approaches to improve performance and foster a climate of continuous innovation and adaptation in the organization.

Srinivas Kilambi • Mahesh Banavar

Large Numerical Models from a Business Perspective

LNM, a Parallel Universe to LLM

Springer

Srinivas Kilambi
Sriya AI
Duluth, GA, USA

Mahesh Banavar
Department of ECE
Clarkson University
Potsdam, NY, USA

ISSN 1933-978X ISSN 1933-9798 (electronic)
Synthesis Lectures on Technology Management & Entrepreneurship
ISBN 978-3-032-14868-1 ISBN 978-3-032-14869-8 (eBook)
https://doi.org/10.1007/978-3-032-14869-8

This Springer imprint is published by the registered company Springer Nature Switzerland AG
The registered company address is: Gewerbestrasse 11, 6330 Cham, Switzerland

To Ceasar, my son, friend, philosopher, and guide. One who has walked, talked, and played with me for last 7 years rain or shine.

To Mushu, for keeping me on the straight and narrow.

Preface

This book details the development and application of the Large Numerical Model (LNM) and its foundational technology, the Sriya Expert Index (SXI). The necessity for this research arose from a significant challenge in modern data science: the need for a highly accurate, reliable, and interpretable framework to manage vast, heterogeneous numerical datasets across critical enterprise domains.

Traditional machine learning (ML) models, while effective for discrete tasks, often struggle with the complexity, scale, and dynamic nature of real-world business data. Furthermore, the recent rise of Large Language Models (LLMs), while revolutionary for text, has highlighted a critical gap in high-precision numerical analysis due to their inherent limitation of hallucination.

The methodology detailed within this text addresses these limitations head-on. The LNM/SXI framework was conceived as an adaptive, domain-agnostic metric that integrates the predictive strengths of multiple ML algorithms and optimizes performance through a proprietary Deep Neural Network.

The case studies presented—covering financial fraud, loan payments, customer churn, supply chain optimization, readmissions, and sepsis prediction—rely on extensive real-world and public domain datasets. These experiments were conducted to validate the framework's superior accuracy and its capacity to deliver actionable, time-phased strategies for quantifiable business improvement.

This book is intended for practitioners, data scientists, and business leaders seeking to transition from single-model predictions to a scalable, universally applicable solution for maximizing positive business outcomes and minimizing risk across their organizations. It is the hope of the authors that this work provides the technical foundation necessary to advance the field of predictive numerical analytics.

Duluth, GA, USA — Srinivas Kilambi
Potsdam, NY, USA — Mahesh Banavar

Acknowledgements

The development of a framework such as this requires team effort. The authors would like to thank Prashant Yadav, Prasoon Jha, Reeshabh Kumar, Deepanshu Gupta, Sanket Kamble, and Mintu Adrak for their contributions.

The authors would also like to thank their spouses, children, friends, and family whose continued guidance, patience, and support enabled completion of this research, development of a truly foundational model, and finally helped in writing and proofreading the book's contents.

Contents

1 Introduction

From mythical automata and mechanical flying machines to the theoretical foundations of computational intelligence, the quest to create artificial life is a story as old as human imagination itself. Ancient texts describe sophisticated mechanical contrivances that, while not autonomous in the modern sense, represented a deep conceptual understanding of advanced, self-moving systems. Centuries later, pioneering thinkers would lay the logical groundwork for the artificial intelligence. This journey from mythological wonder to mathematical precision has culminated in a new era of powerful computational frameworks, including deep networks architectures, leading to large language models and other cutting edge artificial intelligence technologies. In this book, we add this tapestry with Large Numerical Models (LNMs), which leverage advanced algorithms to analyze vast, heterogeneous datasets and provide actionable insights without the limitations of hallucination, bridging the gap between historical fantasy and a future of machine-driven intelligence.

In this introductory chapter, we provide a brief history of machine learning and artificial intelligence, followed by an introduction to the concept of large numerical models. Finally, we preview the rest of the book.

1.1 A Brief History of Machine Learning and Artificial Intelligence

The roots of machine learning and artificial intelligence can be traced back to ancient mythology and literature. Examples include the Golem in Jewish tradition [1], Talos guarding Crete [2], references to "self-moving" aerial vehicles and autonomous guardians in ancient Indian texts [2], and more recently, Mary Shelley's Frankenstein [3]. Deeply

S. Kilambi, M. Banavar, *Large Numerical Models from a Business Perspective*,
Synthesis Lectures on Technology Management & Entrepreneurship,
https://doi.org/10.1007/978-3-032-14869-8_1

embedded in human imagination is an obsession with autonomous systems and artificial intelligence, driving their creation through engineering.

Concurrent with the expression of these fantastic ideas, practical advancements in computation began to emerge. Technologies such as the abacus, Napier's Bones (1617), and the Pascaline (1642) showed progress in the field of mechanical calculators, and the industrial revolution led to Babbage creating the Difference Engine (1822) and the Analytical Engine (1833), the earliest computers [4, 5]. Ada Lovelace was the first computer programmer [6], and Jacquard's Loom demonstrated the first instance of computer memory [7].

The 1940s and 1950s showed the possibility of machine intelligence as a reality. In 1943, neurophysiologist Warren McCulloch and mathematician Walter Pitts published "A Logical Calculus of the Ideas Immanent in Nervous Activity," proposing a mathematical model of the nervous system as a network of simple logical elements, later known as artificial neurons [8]. This seminal work demonstrated that these McCulloch-Pitts neurons could perform all logical functions, thereby providing a computational blueprint for how intelligence could be mechanized, thus laying the foundation for artificial neural networks. This was followed by the work by Turing, and shortly thereafter, by implementations and demonstrations of machine learning, including Arthur Samuel's checkers program. Arthur Samuel first coined the term "machine learning" in his work describing the checkers program [9]. The early adoption of games as a domain for AI research was significant because games provide a simplified, well-defined environment with clear rules, ideal for testing and demonstrating learning and problem-solving algorithms without the complexities of the real world. This approach allowed researchers to focus on core AI challenges like search, heuristics, and learning from experience in a manageable setting.

The formal birth of the field occurred at the 1956 Dartmouth Conference on Artificial Intelligence [10], where the field officially got its name. This was followed by the invention of the Peceptron in 1957 and (M)ADALINE ((Many) Adaptive Linear Element) in 1959, the first neural network applied to real-world problems. The 1970s marked the rise of rule-based expert systems, which were considered early practical AI models [11].

The evolution of computing hardware during this period was instrumental in enabling AI research to progress. The post-World War II era witnessed the transition from electro-mechanical computers to purely electronic ones, initially utilizing vacuum tubes. Machines like the ENIAC (1945) were early electronic programmable computers, significantly increasing computational speed. From the mid-1950s, the replacement of vacuum tubes by transistors led to "second generation" computers that were considerably smaller, consumed less power, and were more reliable. The subsequent "third-generation" computers, emerging in the late 1960s, adopted integrated circuit (IC) chips, making memory cheaper and more power-efficient. These continuous hardware advancements significantly increased computational power and accessibility, providing the necessary infrastructure for more complex AI algorithms to be developed and tested. This hardware evolution was not merely an improvement; it was a fundamental enabler that made AI research feasible on a broader scale, allowing for the transition from "possibilities" to "demonstrations".

Despite the initial optimism and the successes of early expert systems, the 1970s brought a period of significant disillusionment and reduced funding, often referred to as the "first AI Winter", a direct consequence of over-optimistic predictions and the inability of early AI systems to scale from "toy problems" to real-world complexity. This period marked a general decline in interest and investment across both academia and industry. While seemingly a setback, this period forced researchers to confront fundamental limitations and shift focus.

The 1980s saw a renewed interest in AI, primarily fueled by the commercial success of expert systems. These rule-based systems, designed to mimic human experts in specific domains, proved to be highly valuable. Demonstrable economic benefits led to the widespread adoption of expert systems across various industries, including healthcare, and saw a crucial shift from the earlier, more ambitious pursuit of Artificial General Intelligence (AGI) to practical, commercially viable applications.

Alongside the expert systems boom, the 1980s and 1990s witnessed a critical resurgence of connectionist approaches, particularly neural networks, driven by significant algorithmic breakthroughs, including the demonstration of back propagation [12], a fundamental algorithmic breakthrough that made training complex neural networks practical, directly addressing a major technical limitation that had hindered earlier connectionist approaches. This algorithmic innovation, combined with increasing computational power, laid the essential groundwork for the deep learning revolution that would follow decades later. Further advancements included the introduction of Support Vector Machines [13] and Long Short-Term Memory (LSTM) networks [14], seeing the development of a richer toolkit of algorithms, each suited for different types of data and problems, moving ML from a niche area within AI to a distinct and powerful subfield with broader applicability.

While algorithms made advancements, the rise of general-purpose computers in the late 1980s meant the slow-down of the growth of specialized hardware to implement complex machine algorithms. This, coupled with expert systems failing to consistently meet the high expectations set for them, led to another slowdown in the progress of AI. Despite this, the early 1990s still saw some impressive strides in AI research, including the first AI system to beat a reigning world champion chess player, demonstrating that research continued even in challenging climates. 3D graphics cards, precursors to modern GPUs, were also introduced, primarily to enhance video game graphics, laying the foundation for future AI growth.

In 2012, a convolutional neural network (CNN) drastically improved image recognition performance, demonstrating the unprecedented power of deep convolutional networks for large-scale visual recognition tasks [15]. This success was explicitly enabled by the utilization of Graphics Processing Units (GPUs) during training, which provided the massive parallel processing power required for such computationally intensive tasks. CNNs became fundamental for computer vision, while Recurrent Neural Networks (RNNs), particularly those incorporating Long Short-Term Memory (LSTM) units, continued to be critical for processing sequential data, enabling advancements in areas like speech recognition and natural language processing.

A further innovation was the introduction of the Transformer architecture [16], based on the attention mechanism, which fundamentally changed how sequential data was processed by allowing parallel computation rather than relying on recurrence. The Transformer architecture quickly became the underlying basis for most modern Large Language Models (LLMs), revolutionizing natural language understanding and generation.

These successes and the general emphasis on "large amounts of data" signified a profound shift from the knowledge-driven, symbolic AI of previous decades to a data-driven approach. Instead of hand-coding rules, AI systems could learn complex patterns directly from vast datasets, a capability fundamentally enabled by improved computing power and data availability.

The concurrent evolution of cloud computing provided the essential infrastructure for the deep learning revolution. The modern cloud computing era truly began in the 2000s with the emergence of major cloud service providers. Amazon Web Services (AWS) launched cloud services including Simple Storage Service (S3) and Elastic Compute Cloud (EC2), offering on-demand processing, memory, and networking capacity on a pay-as-you-go model. Google followed with App Engine, and later, the Google Cloud Platform (GCP), and Microsoft introduced Azure. These providers rapidly expanded their offerings to include Infrastructure-as-a-Service (IaaS), Platform-as-a-Service (PaaS), and Software-as-a-Service (SaaS) models. This facilitated the availability of Machine Learning as a Service (MLaaS) on the cloud, providing pre-built models, APIs, and tools that simplified the development process for businesses.

The computational demands of deep learning necessitated a new class of hardware accelerators. While Central Processing Units (CPUs) are general-purpose processors efficient for sequential tasks, they are less efficient for the parallel processing required by deep learning. Graphics Processing Units (GPUs), originally designed for rendering complex 2D and 3D images, proved to be exceptionally well-suited for the parallel computations inherent in neural network training. Training deep neural networks on GPUs can be over 10 times faster than on CPUs with equivalent costs. More recently, GPUs have further evolved into highly specialized AI accelerators like Tensor Processing Units (TPUs).

The development of Large Language Models (LLMs) and generative AI has been a defining feature of the current AI landscape. OpenAI's GPT models, particularly GPT-3, demonstrated remarkable text generation capabilities and profoundly impacted natural language understanding. The public release of OpenAI's ChatGPT marked an "overnight success," leading to a global surge in AI usage among consumers, developers, enterprises, and governments. This rapid adoption has led to a proliferation of AI models, both proprietary and open-source, enabling a new generation of products such as advanced chatbots and text-to-image models. These models leverage deep learning techniques but are distinguished by their immense scale in terms of model size and training data, offering high-quality content generation, natural interactions, and versatility across a wide range of tasks.

While the existing LLMs are proficient with content generation and interactions, they are poor when it comes to digesting and processing structured numerical data. This gap is addressed with the advent of large numerical models, or LNMs.

1.2 The Large Numerical Model

Large Numerical Models (LNMs) represent a sophisticated computational framework engineered to process and analyze immense volumes of numerical data with exceptional accuracy. Unlike traditional models, LNMs are pre-trained on extensive datasets, enabling them to deliver reliable "blind" predictions on new, unseen data without the need for extensive retraining. This makes them highly effective for predictive analytics and decision-making across diverse sectors such as healthcare, finance, and supply chain management. A key advantage of LNMs is their ability to avoid "hallucinations"—incorrect or nonsensical outputs—that can be a significant limitation in large language models (LLMs). By integrating technologies like the Sriya Expert Index (SXI) and reinforcement learning, LNMs can harness the power of multiple advanced algorithms to provide robust, scalable, and highly accurate solutions.

The LNM framework is broadly categorized into two types: Homogeneous LNMs (HoLNMs) and Heterogeneous LNMs (HeLNMs). HoLNMs focus on merging datasets from the same or similar domains to create a unified and consistent dataset. This approach enhances the depth of information within a specific area, leading to highly accurate and reliable models for domain-specific applications, such as optimizing chemical formulations. In contrast, HeLNMs integrate data from disparate and adjoining domains, creating master datasets rich with contextual insights. This cross-domain capability allows HeLNMs to capture external influences—like economic trends or environmental factors—and uncover complex interdependencies that would be invisible in isolated datasets, thus enhancing predictive accuracy and enabling more comprehensive, proactive decision-making.

To manage the complexities of integrating diverse and often incomplete datasets, HeLNMs employ advanced deep learning architectures. Generative technologies are used to produce synthetic data and intelligently fill in missing values. This ensures data completeness and robustness without introducing statistical noise. Furthermore, HeLNMs can predict missing target variables, allowing them to function effectively even when complete ground-truth data is unavailable. While facing commercialization hurdles such as scalability and computational efficiency, the integration of LNMs with decision support systems promises to translate complex data into actionable insights, driving significant performance improvements and offering a competitive advantage in a dynamic market.

1.3 Organization of this Book

This book is structured to first introduce the core technologies and then demonstrate their powerful applications through real-world case studies. We begin by detailing the Sriya Expert Index (SXI), which is a dynamic scoring system that converts complex, multi-dimensional problems into a simpler, actionable solution by leveraging a proprietary formula composed of weights from 5–10 machine learning algorithms. This is followed by a discussion of the Large Numerical Model (LNM), an advanced computational framework that integrates and analyzes vast datasets across multiple domains, built upon the SXI++ technology.

The remainder of the book is dedicated to demonstrating the efficacy of the SXI and LNM framework across diverse sectors. These case studies explore critical business and operational challenges, including: predicting and mitigating supply chain disruptions (backorders and shipment delays), enhancing financial risk assessment (loan late payments and fraud detection), minimizing customer churn in the wireless industry, improving marketing analytics for product usefulness, and advancing healthcare predictive modeling (patient readmissions and early sepsis detection). Each chapter highlights the superior accuracy and actionable insights provided by the SXI/LNM approach over traditional machine learning models.

References

1. Idel, M. (1990). *Golem: Jewish magical and mystical traditions on the artificial anthropoid*. State University of New York Press.
2. Mayor, A. (2018). *Gods and robots: Myths, machines, and ancient dreams of technology*. Princeton University Press.
3. Shelley, M. (1818). *Frankenstein; or, the modern Prometheus*. Lackington, Hughes, Harding, Mavor, & Jones.
4. Felt, D. E. (1916). *Mechanical arithmetic, or the history of the counting machine*. Washington Institute.
5. Roegel, D. (2015) *Napier's bones and Genaille-Lucas's rods*. LORIA, Nancy, Technical Report.
6. Fuegi, J., & Francis, J. (2003). Lovelace & Babbage and the creation of the 1843 'Notes'. *IEEE Annals of the History of Computing, 25*(4), 16–26.
7. Anderson, D., & Delve, J. (2007). Biographies [F.C. Williams; J. Vaucanson; J.M. Jacquard]. *IEEE Annals of the History of Computing, 29*(4), 90–102.
8. McCulloch, W. S., & Pitts, W. (1943). A logical calculus of the ideas immanent in nervous activity. *The Bulletin of Mathematical Biophysics, 5*(4), 115–133.
9. Samuel, A. L. (1959). Some studies in machine learning using the game of checkers. *IBM Journal of Research and Development, 3*(3), 210–229.
10. Dartmouth College. (n.d.) *Artificial intelligence (AI) coined at Dartmouth* [Online]. Retrieved July 31, 2025, from https://home.dartmouth.edu/about/artificial-intelligence-ai-coined-dartmouth
11. Foote, K. D. (2022). The history of machine learning and its convergent trajectory towards AI. In *Machine learning and the city: Applications in architecture and urban design* (pp. 129–142). Wiley.
12. Rumelhart, D. E., Hinton, G. E., & Williams, R. J. (1986). Learning representations by back-propagating errors. *Nature, 323*(6088), 533–536.
13. Cortes, C., & Vapnik, V. (1995). Support-vector networks. *Machine Learning, 20*(3), 273–297.
14. Hochreiter, S., & Schmidhuber, J. (1997). Long short-term memory. *Neural Computation, 9*(8), 1735–1780.

15. Krizhevsky, A., Sutskever, I., & Hinton, G. E. (2012). ImageNet classification with deep convolutional neural networks. *Advances in neural information processing systems, 25*, 1097–1105.
16. Vaswani, A., et al. (2017). Attention is all you need. *Advances in Neural Information Processing Systems, 30.*

Introducing Sriya Expert Index (SXI): The Technology that Powers the LNM

2

SXI is a dynamic score/index obtained from a proprietary formula consisting of weights from 5–10 ML algorithms [1]. SXI is a super feature and is a true weighted representative of all important features which converts a multi-dimensional hard-to-solve problem into a simpler 2-dimensional solution. SXI with Proprietary Deep Neural Network algorithm involves an iterative approach where it dynamically adjusts algorithm weights based on the most significant weights provided by the 5–10 ML algorithms. This process aims to enhance the correlation between SXI scores and business outcomes, thereby improving accuracy and delineation.

2.1 SXI Important Features Evaluation

The SXI Score or SXI Index, is a comprehensive set of algorithms designed to evaluate and quantify the business experience. As a dynamic system, the SXI employs machine learning techniques to assign individual learning weights to the top five features based on their significance and impact on the learning process. Each algorithm autonomously identifies and evaluates the most influential features within the data, considering multiple factors within the data. By assigning learning weights to these top features, the SXI ensures an iterative and adaptive approach, allowing the system to prioritize and weigh the importance of specific aspects in shaping the overall index score. This personalized weighting mechanism enables the SXI to dynamically adjust its evaluation criteria.

The SXI Framework (Fig. 2.1) employs a variety of mathematical techniques ranging from arithmetic to statistical analyses and non-machine learning algorithms to process cleaned data effectively. By leveraging this diverse framework, the SXI utilizes multiple functions, including statistical measures like Standard Deviation and various types of

S. Kilambi, M. Banavar, *Large Numerical Models from a Business Perspective*,
Synthesis Lectures on Technology Management & Entrepreneurship,
https://doi.org/10.1007/978-3-032-14869-8_2

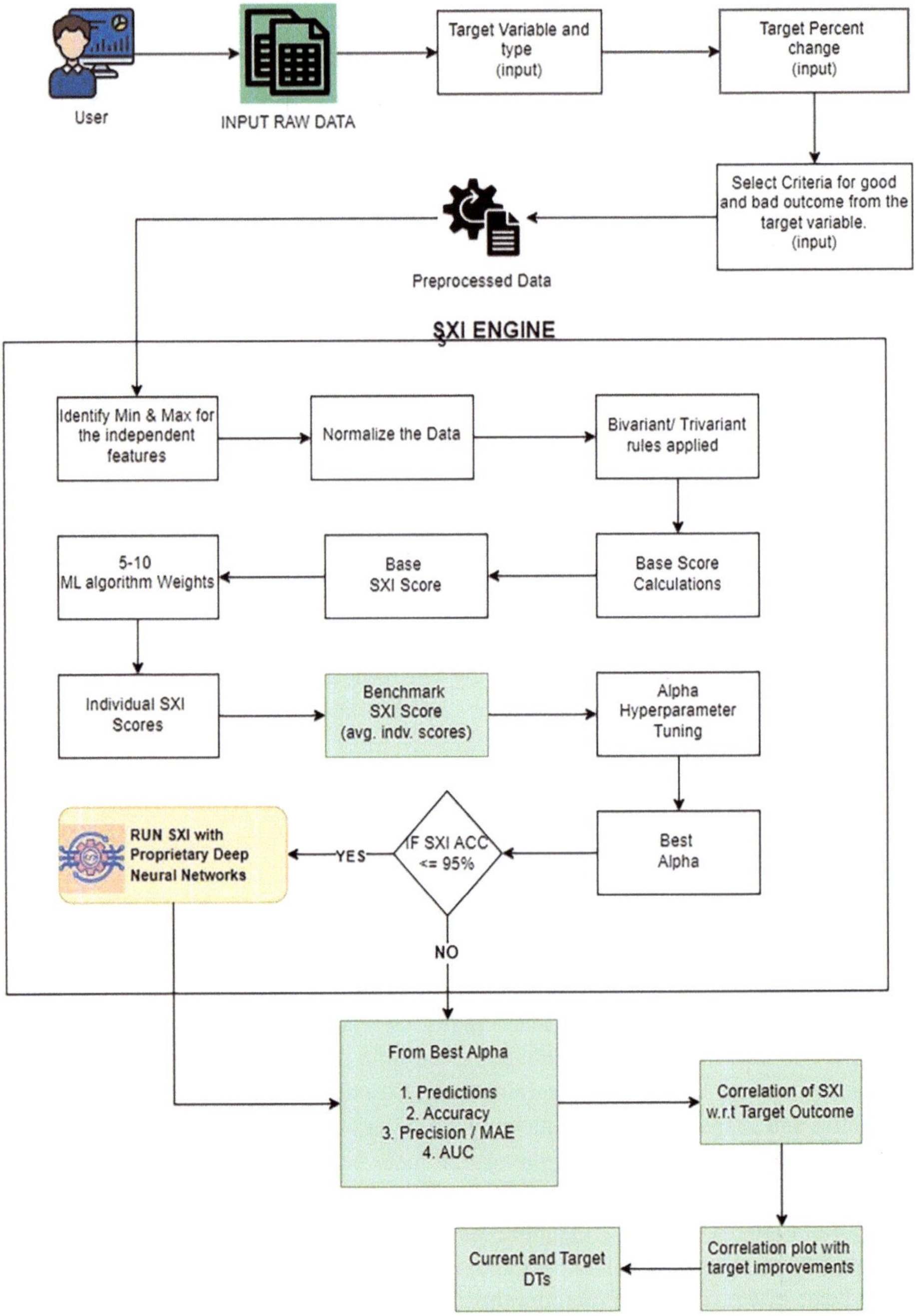

Fig. 2.1 SXI framework

correlations, alongside algebraic operations such as proportionality constants. Through these operations, the SXI Engine calculates a Base SXI score by bringing together various statistical analysis and algebraic computations.

The SXI Engine utilizes algebraic and statistical methods to derive weights for individual user input parameters within a user vector. These weights act as multipliers, enabling adjustments to the significance of each parameter. For example, bi-variate and multi-variate analyses may yield correlation coefficients, which are then employed as multipliers for the user vector components. This process contributes to the generation of a robust Base SXI score, serving as the foundation for subsequent steps in the SXI algorithm.

Once the Base SXI score is established, the system incorporates weights obtained from 5 to 10 machine learning algorithms. These weights collectively contribute to form the Final SXI scores, which serve as the Initial SXI or the Benchmark SXI score in the SXI algorithm. This integration of diverse mathematical functions, statistical analyses, and machine learning algorithms highlights the versatility of the SXI, providing a comprehensive approach to scoring and delineating business outcomes based on user input parameters.

Benchmark analysis of the baseline score involves extracting key metrics such as the Average SXI Score (Benchmark Score), percentage of good and bad outcomes, and the distribution of outcomes above and below the average SXI. This analysis offers insights into the performance of the SXI algorithm in distinguishing between positive and negative outcomes.

The Proprietary Deep Neural Network algorithm (Fig. 2.2) plays a crucial role in optimizing the SXI algorithm. It adjusts weights dynamically based on the performance of the system, aiming to enhance the correlation between SXI scores and business outcomes. The Proprietary Deep Neural Network algorithm agent identifies the top features from machine learning weights and adjusts the weights accordingly. Additionally, it introduces a custom weight initialization strategy based on feature importance, aligning the weight initialization process with the significance of individual features [2, 3].

The Deep Learning Architecture consists of automated hyperparameter tuning, flexibility in activation functions and optimizers, dynamic architecture based on dataset size, and structured evaluation metrics. These characteristics ensure adaptability, efficiency, and thorough evaluation of the deep neural network model.

The workflow of the Proprietary Deep Neural Network algorithm involves steps such as train-test split, in our study, we divided the dataset into 70% for training, 20% for testing, and 10% for validation. The validation set, making up 10% of the data, plays a crucial role in the design of SXI algorithm by aiding in hyperparameter tuning, overfitting detection, and appropriate model selection, ensuring that the model can generalize well to new, unseen data. Overfitting, where a model performs well on training data but poorly on real-world data, can be mitigated with the validation set. This thorough process ensures that the model is not only accurate on the training data but also reliable and generalizable. The validation dataset is thus a critical stage in the SXI workflow, confirming that the model remains highly accurate and generalizable when applied to new data. In this model, we adopted a 70-20-10 split for training, testing, and validation. This approach ensures that

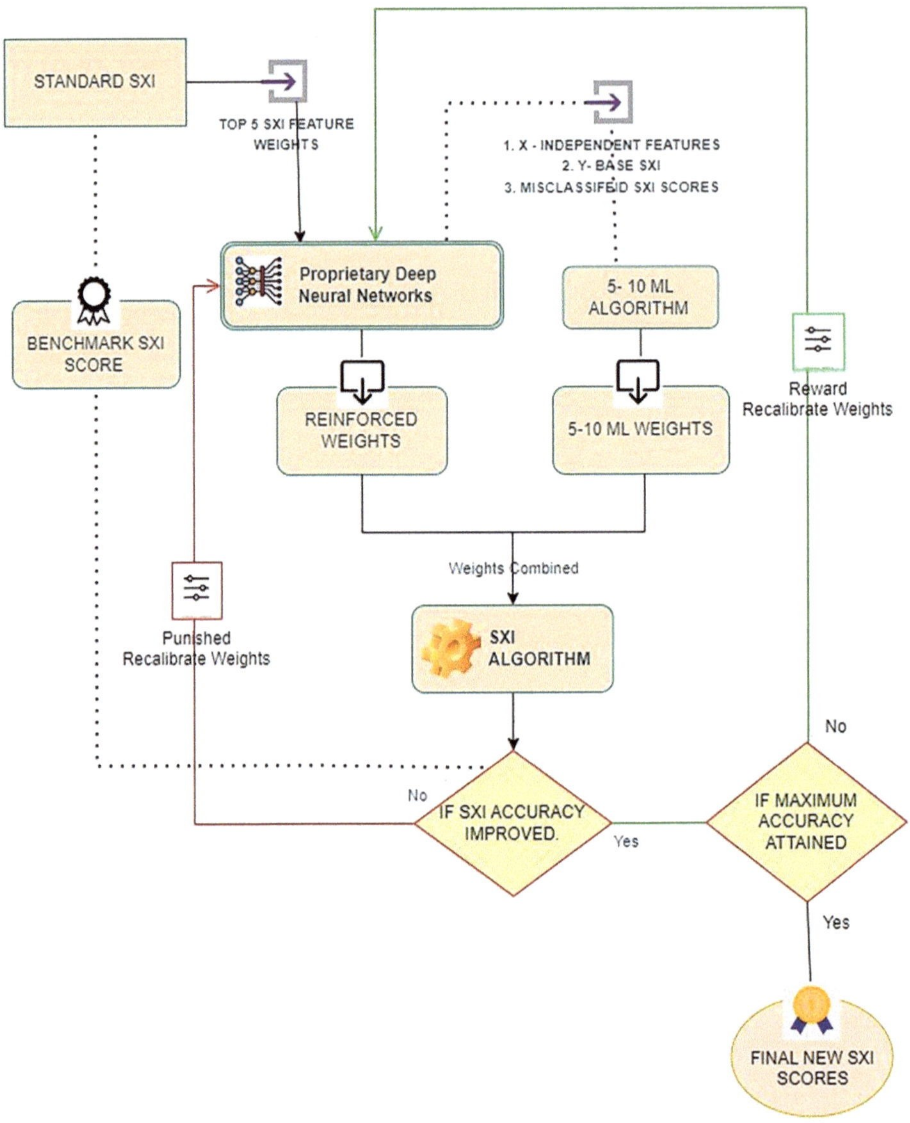

Fig. 2.2 Proprietary DNN architecture

the model is thoroughly evaluated and fine-tuned before being tested on the 20% of data set aside for final performance assessment. The testing set, which comprises 20% of the data, is used to evaluate the model's performance and ensure its accuracy on new, unseen data.

Bayesian optimization for hyperparameter tuning, model configuration, and model training and evaluation with the best hyperparameters. Iterative weight calibration is performed to refine the SXI algorithm, where it adjusts weights, applies rewards and punishments based on performance improvements, until an optimal delineator is achieved.

In the reward and punishment mechanism, improvements in both SXI score and class delineation accuracy lead to a reward. This involves integrating the newly adjusted weights into the existing ones and recalculating SXI scores to evaluate the enhancements in delineation and accuracy. If these improvements occur without any weight adjustment, there is a positive weight adjustment, initially ranging from 0 to 100% percentage increase weights. Further iterations iteratively adjust weights until maximum accuracy is achieved. If positive adjustments fail to surpass the initial class delineation, negative weight adjustments from 0 to −100% percentage increase weights are implemented. If these adjustments yield improved delineation and accuracy, subsequent iterations continue adjusting feature weights negatively until no further improvement is possible. In cases where no improvement is observed for both positive and negative adjustments, the reinforced function penalizes by providing the next set of weights, along with additional weightage for the top five most important features in the hidden layers of the deep neural network.

The iterative process continues until the SXI scores cannot be further improved. The final new SXI scores are then updated to the system for further analysis and decision-making regarding business outcomes.

2.2 SXI Correlation with Target Business Outcome(s)

Correlation is a statistical measure that quantifies the extent to which two variables change together. It indicates the direction (positive or negative) and strength of the linear or nonlinear relationship between two variables. A positive correlation suggests that as one variable increases, the other tends to increase as well. In contrast, a negative correlation implies that as one variable increases, the other tends to decrease. A correlation of 0 indicates no linear relationship.

SXI scores and target business outcomes are correlated by employing polynomial regression to identify the best fit curve. This method involves fitting a polynomial equation to the data, allowing us to capture the intricate relationships between SXI scores and business outcomes. By utilizing polynomial regression, we aim to create a curve that not only accurately represents the observed data points but also provides insights into the underlying trends and patterns. This enables us to make informed predictions about the potential impact of SXI score variations in the target business outcomes. The utilization of polynomial regression enhances our ability to model the complex dynamics between business experiences, as reflected by SXI scores, and the desired business outcomes, fostering better strategic decision-making and optimization efforts.

2.3 R-Squared (Coefficient of Determination)

R-squared is a statistical measure that represents the proportion of the variance in the dependent variable that is explained by the independent variable(s) in a regression model. It ranges from 0 to 1, or 0% to 100%. It is easier for humans to view graphical outputs of numerical data, rather than lists of numbers. Thus, SXI invention includes its own graphical user interface for displaying emerging patterns. Examples of such graphical output are simple graphs, bar charts, and more complex outputs such as fitted curves and decision trees. The present product also includes its own user interface allowing the necessary components for creating lists of inputs to be used, adding or deleting inputs from the lists, and other operations for controlling the performance of the SXI Engine and SXI Engine, as well as the various graphical outputs.

A further advantage is that the SXI Engine operates as an ensemble, and the components of these ensembles can be easily changed are new algorithms become available, or current algorithms are found not to provide utility. The SXI score, calculated mathematically based on various input parameters, serves as a numerical representation of business experience. In the context of business outcomes, this score can be correlated with specific classes such as "good" and "bad" within the dataset. One common metric for evaluating the strength of this correlation is the coefficient of determination, often denoted as R-squared (R^2). R-squared quantifies the proportion of the variance in the dependent variable (in this case, business outcomes) that can be explained by the independent variable (SXI score). For example, if a high R-squared value is obtained, it indicates that the SXI score has a strong explanatory power for predicting business outcomes. Conversely, a low R-squared value suggests a weaker correlation. An R-squared value of 1 indicates a perfect correlation, while a value of 0 signifies no correlation.

If the SXI scores (x) are the independent variable and the target outcome (y) is the dependent variable, a polynomial regression equation can be expressed as:

$$y = b_0 + b_1 x + b_2 x^2 + \ldots + b_n x^n + \in$$

Where:

- y is the target outcome.
- x is the SXI scores.
- $b_0, b_1, b_2, \ldots, b_n$ are the coefficients representing the weights of the polynomial terms.
- n is the degree of the polynomial.
- $\in$ is the error term.

The objective of polynomial regression is to estimate the coefficients $b_0, b_1, b_2, \ldots, b_n$ in a way that minimizes the sum of squared differences between the observed values of y and the values predicted by the polynomial equation.

The mathematical expression for the sum of squared errors (SSE) would be:

$$\text{SSE} = \sum_{I=1}^{N} \left(y_i - \left(b_0 + b_1 x_i + b_2 x_i^2 + \ldots + b_n x_i^n \right) \right)^2$$

where N is the number of data points, y_i is the observed value of the target outcome for the i-th data point, and x_i is the SXI score for the i-th data point.

The coefficient of determination often denoted as R^2 is a measure of how well the dependent variable (the target outcome y) is explained by the independent variable(s) (the SXI scores x) in a polynomial regression model.

The R^2 value is calculated using the formula:

$$R^2 = 1 - \left(\frac{\text{SSE}}{\text{SST}}\right)$$

Where:

- SSE is the sum of squared errors, which represents the difference between the observed values of the dependent variable and the values predicted by the regression model.
- SST is the total sum of squares, which represents the total variance in the dependent variable.

The total sum of squares (SST) is calculated as:

$$\text{SST} = \sum_{I=1}^{N} \left(y_i - \overline{y}\right)$$

Where:

- N is the number of data points.
- y_i is the observed value of the dependent variable for the i-th data point.
- $\overline{y}$ is the mean of the observed values of the dependent variable.

When analysing a polynomial regression plot with x (SXI scores) and y (business outcome), one can identify different types of improvements based on the behaviour of the regression curve.

The SXI Score serves as a crucial metric correlated with business outcomes, offering valuable insights. In the initial stages, businesses implementing the SXI framework may witness a noticeable score improvement, reflecting immediate adjustments and optimizations based on early analyses. This short-term enhancement might result from swift corrective actions addressing identified pain points or inefficiencies in the business experience.

Moving into the mid-term, the SXI score may experience a more substantial improvement as businesses continue to refine their strategies and incorporate iterative changes. This phase involves a deeper understanding of user interactions, allowing for more targeted interventions to enhance specific features or aspects contributing to a positive business experience.

In the long term, sustained efforts guided by SXI can lead to a significant and consistent improvement in business outcomes. Continuous refinements based on the evolving digital landscape, user preferences, and business goals contribute to a mature and optimized business experience. The long-term SXI score improvement reflects an organization's commitment to adapting, ultimately translating into positive and enduring impacts on overall business performance.

In practical terms, consider a scenario where a company uses the SXI score to evaluate user interactions on their website. A high R-squared value might suggest that a higher SXI score is associated with positive outcomes, such as increased customer satisfaction, higher conversion rates, or longer customer retention. Conversely, a lower SXI score might align with negative outcomes, such as higher bounce rates or reduced customer engagement. It's essential to interpret the R-squared value in conjunction with domain-specific knowledge and contextual understanding to draw meaningful insights about how the SXI score correlates with specific business outcomes in the dataset.

The correlation between SXI scores and business outcomes not only provides insights into the present state of business experiences but also holds the potential for guiding future improvements. By understanding how the SXI score aligns with business outcomes, organizations can implement targeted strategies to enhance both the SXI scores and the associated outcomes. For instance, if a high SXI score correlates positively with positive business outcomes, the organization can identify the contributing factors that lead to such success. This insight can be used to inform future improvements in business experiences, allowing the refinement of specific features or processes that contribute to a positive business outcome. Moreover, the correlation analysis can guide the allocation of resources towards optimizing those aspects of the business that have a substantial impact on desired business outcomes. Implementing changes based on this correlation can lead to an iterative process of refinement, where adjustments are made to enhance the business experience, consequently positively impacting business outcomes over time.

In summary, leveraging the correlation between SXI scores and business outcomes not only aids in understanding the current scenario but also serves as a strategic tool for continuous improvement, fostering a cycle of refinement that benefits both the SXI scores and the overall business performance in the future.

2.4 Model Training and Evaluation

The model training process starts with hyperparameter tuning, specifically focusing on the alpha parameter. This involves adjusting the SXI score by varying the alpha value between 0.5 and 1.5 in increments of 0.1. During each iteration, the SXI score is recalculated by multiplying the current SXI score with the alpha value, enabling fine-tuning to optimize the model's performance.

$$SXI_{new} = \propto * SXI_{Current},$$

Where:

- $\propto$ is the tuning parameter, ranging from 0.5 to 1.5 in increments of 0.1,

- $SXI_{Current}$ is the Benchmark SXI score, and
- SXI_{new} is the new SXI score after applying the alpha adjustment.

The dataset is divided into three parts: 70% for training, 20% for testing, and 10% for validation. This strategic split ensures the model is well-trained while also providing sufficient data for testing and validation, reducing the risk of overfitting and enabling an accurate assessment of model performance. During training, the optimized SXI scores are used as a "Super Feature," playing a crucial role in enhancing the model's predictive accuracy.

To thoroughly evaluate the model's performance, various metrics are employed. Accuracy measures the overall correctness of the model's predictions, while precision focuses on the proportion of true positive results among all positive predictions. The confusion matrix offers a detailed breakdown of true positive, true negative, false positive, and false negative outcomes, providing deeper insights into the model's classification abilities. Additionally, the Area Under the Curve (AUC) is utilized to assess the model's effectiveness in distinguishing between classes, serving as a robust indicator of its predictive power. Accuracy and precision are defined as:

$$\text{Accuracy} = \frac{\text{TP} + \text{TN}}{\text{TP} + \text{TN} + \text{FP} + \text{FN}},$$

$$\text{Precision} = \frac{\text{TP}}{\text{TP} + \text{FP}},$$

Where:

- TP is the number of true positives,
- TN is the number of true negatives,
- FP is the number of false positives, and
- FN is the number of false negatives.

The AUC for the Receiver Operating Characteristic (ROC) curve measures the classifier's ability to distinguish between positive and negative classes. It is calculated based on the True Positive Rate (TPR) and False Positive Rate (FPR), with the ROC curve plotting TPR against FPR at different threshold settings. TPR, FPR and AUC are defined as follows:

$$\text{TPR} = \frac{\text{TP}}{\text{TP} + \text{FN}},$$

$$\text{FPR} = \frac{\text{FP}}{\text{FP} + \text{TN}},$$

$$\text{AUC} \approx \sum_{i=1}^{n-1} \left(\frac{\text{TPR}_{i+1} + \text{TPR}_i}{2} \right) \left(\text{FPR}_{i+1} - \text{FPR}_i \right),$$

Where:

- TPR_i and TPR_{i+1} are the true positive rates at consecutive thresholds, and
- FPR_i and FPR_{i+1} are the false positive rates at consecutive thresholds.

2.5 SXI Decision Trees (The Final Output)

In the present product version, the Cleaned Data and the SXI scores are presented to a Decision Tree Algorithm. The Decision Tree (DT) algorithm classifies input data (users) by some specified criteria or class. In the simplest form of this embodiment, the classification requested of the DT algorithm is to group users as having "good" or "bad" business outcome experience as indicated by having an SXI score above or below the average SXI score for the entire dataset presented to the DT algorithm. The DT algorithm generates the necessary nodes, which are the individual user input parameters, such as "Sample value %," "Promo value," "headquarter name," etc. (Note these inputs will differ with different data). Further, the DT algorithm also determines the numerical value for each individual user input parameter above or below which the population is divided. Categorical values may be handled as dummy variables, having numerical values of 1 or 0, for "in class"/"not in class," respectively. The nodes of the resulting decision tree can be ranked in order of "split strength," that is, the degree to which they correctly divide the data at that point in the tree between the classes. This ranking of nodes gives the relative importance of the individual user input parameters for performing the classification, and thus for the user's experience. This allows the SXI Engine to look up and present possible corrective actions for individual user input parameters that have high split strength and partition many data points into the "bad experience" class.

Another embodiment, iterative calculations are performed using the SXI Algorithm. In the iterative process, changes to the SXI Algorithm may be provided by manual instructions provided by a human operator, instructions from ML algorithms in the SXI Engine, or a combination of both. Iteration would continue until a suitably small change in the overall SXI score occurs from one iteration to the next. In the context of the described decision tree algorithm, the auto-generated nodes can be interpreted by both artificial intelligence systems, specifically the Gen AI components or AI Agents, and by human experts. The interpretability of these nodes ensures transparency in understanding the decision-making process. Gen AI, equipped with sophisticated algorithms, can autonomously analyze and derive insights from the structured hierarchy of nodes. Meanwhile, human analysts have the flexibility to manually review and interpret the decision tree nodes, gaining valuable insights into the factors influencing the classification.

This dual interpretability allows for a collaborative and comprehensive approach to decision-making. While advanced AI systems like Gen AI can expedite large-scale analyses, human interpretability ensures that domain expertise and understanding can be applied to validate and refine the insights generated by the automated decision tree. This symbiotic relationship between machine intelligence and human interpretation fosters a more robust and accountable decision-making process in the context of the presented auto-tree nodes. Interpreting decision tree nodes plays a pivotal role in enhancing target business outcomes. Understanding the significance of each node allows businesses to pinpoint the specific factors influencing the classification of user experiences as "good" or "bad." This insight enables targeted interventions and strategic optimizations that can positively impact key performance indicators and overall business success.

For instance, if the interpretation reveals that certain user input parameters, such as "fast delivery" or "free promo gifts," strongly contribute to a positive user experience, businesses can focus on improving these aspects. This targeted approach facilitates the development of personalized strategies to increase positive customer feedback or encourage more frequent purchases, ultimately enhancing customer satisfaction. Conversely, if nodes indicate a negative impact on user experiences, businesses can prioritize corrective actions to address these pain points. For example, understanding that a high product returns or poor product feedback correlates with a "bad" experience may prompt efforts to offer incentives, or improve product information. The iterative process of interpreting and acting upon decision tree nodes allows businesses to refine their digital strategies continually. This agility in response to user behavior patterns, as revealed by the decision tree, aligns with evolving customer expectations, potentially resulting in overall improved business outcomes.

Achieving a targeted $x\%$ improvement in a specific business outcome involves a strategic and comprehensive approach. To bring about this improvement, $x\%$ changes are systematically applied to the entire set of features within the dataset, irrespective of whether these features carry positive or negative weights. This inclusive adjustment allows for a balanced optimization process, addressing both strengths and weaknesses identified through data analysis. Positive-weighted features, indicating factors associated with favorable outcomes, may be enhanced or further leveraged, while negative-weighted features, pointing to aspects affecting the outcome negatively, can be mitigated or reconfigured. By implementing a percentage change uniformly across the feature set, organizations ensure a holistic and data-driven effort to attain the desired improvement in the target business outcome, fostering a more effective optimization strategy.

2.6 Actionable Insights from a Decision Tree for Improving Business Outcomes (Late Payments Dataset as an Example)

In the decision tree model, borrower and financial behavior features contributing to late payments are assigned positive weights, while those mitigating late payments are assigned negative weights. To implement strategies for reducing late payments, users specify a percentage increase for positive weighted features and a percentage decrease for negative weighted features. During the data transformation process, these adjustments are applied to each observation: positive weighted features are increased by the specified percentage, and negative weighted features are decreased, simulating an initial enhancement in the dataset to reduce late payment risks [4].

If the late payment risk model shows a positive correlation between SXI scores and late payment risk, the strategy for positive weighted features is:

$$\text{Adjusted Feature Value} = \text{Orginal Feature Values} * \left(1 + \text{Percentage Increase}\right).$$

For features that reduce late payment risk, the strategy is:

$$\text{Adjusted Feature Value} = \text{Orginal Feature Values} * \left(1 - \text{Percentage Decrease}\right).$$

In cases where a negative correlation exists between SXI scores and late payment risk (where higher SXI scores indicate lower late payment risks), the weights are reversed. Features negatively impacting late payment risks (increasing them) are given positive weights, while features reducing late payment risks receive negative weights. Under this scenario, positive weighted features (which increase late payment risk) require a strategy involving a percentage decrease, and negative weighted features (which reduce late payment risk) call for a percentage increase.

For positive weighted features, the strategy is:

$$\text{Adjusted Feature Value} = \text{Orginal Feature Values} * \left(1 - \text{Percentage Decrease}\right).$$

For negative weighted features the strategy is:

$$\text{Adjusted Feature Value} = \text{Orginal Feature Values} * \left(1 + \text{Percentage Increase}\right).$$

Once the dataset has been adjusted using these transformations, it is used to train a Random Forest model. The model learns the relationships between borrower attributes, financial behaviours, and late payment risks by analysing data through multiple decision trees. Each decision tree captures distinct aspects of the feature-risk relationship, and among these, a target decision tree path is identified. This path represents the sequence of feature splits that most accurately predicts late payment risks based on the adjusted dataset, offering a detailed understanding of the factors contributing to late payments and actionable insights into effective mitigation strategies.

References

1. Kilambi, S. (2024). *AI square enabled by Sriya Expert Index (SXI): Method of determining and use*. U.S. provisional patent application no. 63/549,553, 4 Feb 2024.
2. Kilambi, S. (2024). *AI square enabled by Sriya Expert Index (SXI): Plus reinforcement learning*. U.S. provisional patent application no. 63/553,335, 14 Feb 2024.
3. Kilambi, S. (2024). *AI square enabled by Sriya Expert Index (SXI): Generative AI*. U.S. provisional patent application no. 63/554,252, 16 Feb 2024.
4. Kilambi, S. (2024). *Processing of large numerical models (LNM) by AI2 enabled SXI*. U.S. provisional patent application no. 63/575,991, 8 Apr 2024.

3 Large Numerical Models (LNMs): Theory and Practice

Large Numerical Models (LNMs) are advanced computational frameworks designed to process vast amounts of numerical data with high accuracy. Unlike traditional models, LNMs integrate and analyse large datasets across multiple domains, generating precise and actionable insights. These models are pre-trained using SXI++ on extensive numerical datasets, ensuring robust performance in predictive analytics and decision-making.

LNMs offer several advantages over standard and even advanced machine learning models. They exhibit high accuracy and precision while ensuring no hallucinations, a key limitation in large language models (LLMs). Additionally, LNMs are pre-trained and validated, providing reliable "blind" predictions on "unseen data" without requiring extensive retraining or validation. Large Number Models (LNM) encompass a category of mathematical models tailored to handling computations involving exceptionally large numbers, often surpassing the capacities of standard computing systems.

These models find crucial applications across diverse sectors including healthcare, finance, supply chain, and marketing. In healthcare, LNMs aid in analyzing extensive datasets for medical research and drug discovery. In finance, they facilitate complex financial modeling and risk assessment. LNMs are also instrumental in optimizing supply chain operations and forecasting demand in marketing.

These models leverage specialized algorithms and data structures optimized for efficiency and scalability, enabling researchers and professionals to tackle intricate computational challenges and derive valuable insights across various domains. Large Number Models (LNM) incorporate SXI (Sriya Expert Index) technology with reinforcement learning and LLM based Gen AI as add on technologies, which harnesses the power of 5–10 advanced algorithms. These technologies enable LNMs to process and analyze vast datasets efficiently, providing valuable insights and solutions across diverse domains.

S. Kilambi, M. Banavar, *Large Numerical Models from a Business Perspective*,
Synthesis Lectures on Technology Management & Entrepreneurship,
https://doi.org/10.1007/978-3-032-14869-8_3

By leveraging multiple algorithms simultaneously, SXI technology enhances the accuracy, robustness, and scalability of LNMs, making them invaluable tools for addressing complex computational challenges in multiple fields such as healthcare, finance, supply chain management, marketing, etc.

3.1 Method

The concept of Large Number Models (LNM) represents a significant advancement in AI technology, particularly in its ability to amalgamate the collective knowledge and insights derived from individual SXI models. By integrating the learnings from diverse SXI models across various verticals and use cases, LNMs become exceptionally robust and powerful Model. The strength of LNMs lies in their capacity to learn from a wide range of business data originating from different industries and sectors. These models can ingest and process vast quantities of data from diverse sources. By leveraging the collective intelligence encoded within the SXI models, LNMs gain a comprehensive understanding of complex business dynamics and patterns across multiple domains.

Furthermore, the integrated nature of LNMs allows them to recognize and adapt to patterns and trends that may transcend individual industries or use cases. For example, insights gained from analyzing customer behavior in the retail sector may inform strategies for optimizing patient care pathways in healthcare or enhancing supply chain efficiency in manufacturing. SXI LNM would provide a seamless and intuitive experience for users, enabling them to access industry-specific use cases tailored to their needs. Upon accessing the LNM platform, users are presented with a selection of industries and use cases to choose from.

Once the user selects a specific industry and use case, the SXI results, and the underlying data used for generating those results are showcased in a clear and comprehensive manner. The results will include visualizations, detailed reports that highlight key insights, trends relevant to the selected industry and use case.

By dividing the SXI LNM into industry-specific use cases (Fig. 3.1), users can easily navigate through relevant information and gain actionable insights specific to their domain. This approach ensures that users can quickly access the insights they need to make informed decisions and drive business outcomes effectively.

Additionally, showcasing the underlying data used for SXI results enhances transparency and trust, allowing users to validate the findings and understand the basis for the recommendations provided by SXI LNM. Finally, LLNs will be integrated with LLMs (large language models) so all type of data (structured and unstructured) can be fed into the SXI technology suite providing solutions far superior to any solutions available and will potentially be the closest algorithm or technology suite to Artificial General Intelligence (AGI).

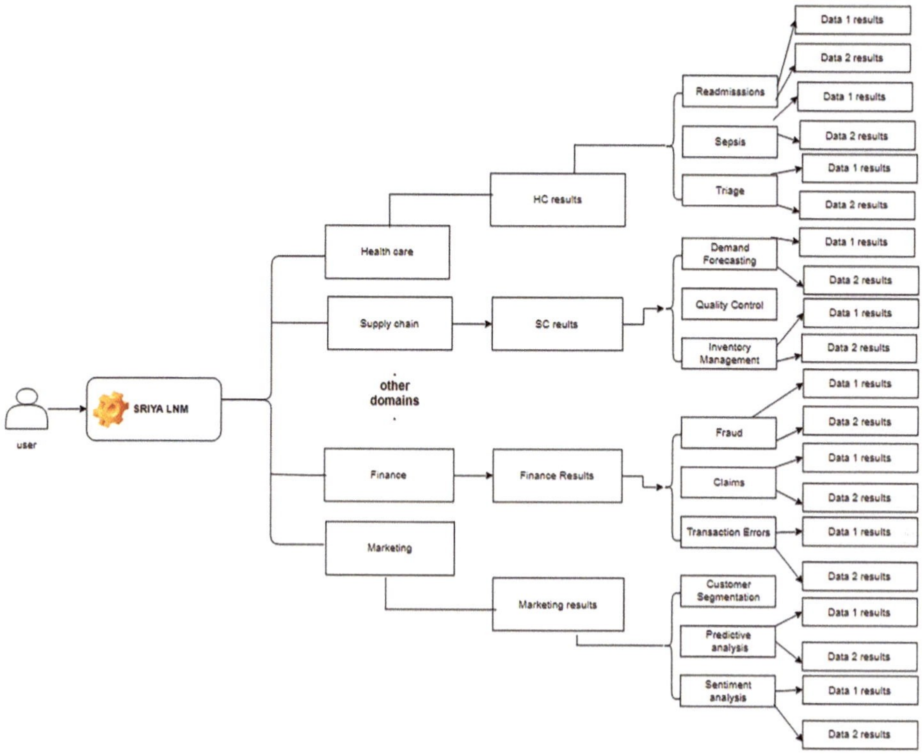

Fig. 3.1 Example industry-specific applications of the SXI LNM

3.2 Types of LNMs

3.2.1 Homogenous LNM

HoLNMs powered by SXI++ specialize in merging multiple datasets from the same or similar domain, creating a unified dataset that enhances consistency and depth of information. These models focus on domain-specific integration, ensuring a focused scope that improves accuracy and reliability. By maintaining consistency and uniformity, HoLNMs simplify data processing and lead to more stable predictive models. Their structured approach makes them ideal for applications such as biomass-lignin-based cement formulation, where precise chemical composition, reactivity, and curing need to be optimized.

HoLNMs are highly effective in structured environments where data originates from related sources, ensuring improved model accuracy and simplified data integration. However, their key limitation is their inability to incorporate multiple adjoining datasets from diverse domains, restricting their ability to capture complex interdependencies. This limitation introduces the need for Heterogeneous LNMs (HeLNMs), which integrate datasets from different domains to create master datasets with richer contextual insights.

HoLNMs merge multiple datasets from the same or similar domain or type. Creates a unified dataset that enhances consistency and dept. of information within a specific area. Focusing on improving accuracy and reliability within a defined scope.

Characteristics of HoLNMs:

- **High Accuracy and Precision:** LNM boasts >95% accuracy and precision, minimizing errors and ensuring reliable Predict.
- **No Hallucinations:** Unlike some AI models, LNM does not produce hallucinations, which are incorrect or nonsensical outputs generated by the model.
- **Pre-trained and Validated:** LNMs are pre-trained on billions of numerical data points, making them robust and reliable without the need for additional spot training.
- **Scalability and Distribution:** LNM can follow a distributed architecture, with domain-wise segmentation for different use cases, enhancing processing speed and efficiency.
- **Data Integration and Feature Analysis:** LNMs integrate top features from individual models within a domain to train a master model, improving prediction accuracy.

Two Phases of HoLNMs Operation:

- **Past Data Exists:** When past data is available, it is used to train new sub-models within the domain, and the new data's contribution is checked against the master model. This phase allows for continuous improvement and feature identification.
- **No Past Data Exists:** In cases where no past data exists, the system relies on common features identified from previously executed models within the same domain to guide new data capture.

3.2.2 Heterogenous LNM (HeLNM)

Heterogeneous Large Numerical Models (HeLNMs) represent a transformative approach to data-driven modelling, enabling the integration of multiple datasets from adjoining domains. Unlike Homogeneous Large Numerical Models (HoLNMs), which focus on structured data from similar domains, HeLNMs merge disparate datasets to create a master dataset with richer contextual information and improved training potential. This cross-domain approach enhances predictive accuracy and decision-making by capturing external influences that significantly impact outcomes not possible in ML, LLMs or even HoLNMs.

A defining advantage of HeLNMs is cross-domain integration, where data from distinct fields—such as material science, supply chain, energy production, and environmental factors—are combined. This integration provides an enhanced data context, ensuring that interdependencies between variables across different domains are captured effectively. By merging heterogeneous data, these models uncover deeper patterns and insights that remain hidden in isolated datasets, making them particularly valuable for complex industrial and business applications. Another advantage of HeLNMs is their increased training potential. HeLNMs, powered by the SXI++ algorithm leverage a robust data fusion approach, enabling the model to learn from a broader range of inputs. This results in more accurate predictions, improved adaptability, and greater applicability to real-world industrial challenges.

The ability of HeLNMs to capture complex interdependencies is particularly valuable in industries where multiple variables interact simultaneously. For example, in industrial material optimization, factors such as raw material composition, processing conditions, and supply chain constraints can be influenced by external variables like market demand, weather conditions, or energy prices. By incorporating these external factors, HeLNMs provide a deeper level of analysis, ensuring that businesses gain more actionable and valuable insights tailored to dynamic real-world conditions. The most significant advantage of HeLNMs is their ability to ensure external influences. Such as economic trends, regulatory changes, or environmental conditions, are considered, allowing businesses to make proactive, data-driven decisions with greater confidence.

3.2.2.1 Flexible Predictive Ability of HeLNMs on Unseen Data

HeLNMs powered by SXI++ enable flexibility in prediction without requiring retraining on new data. By leveraging a pre-trained deep learning architecture, HeLNMs dynamically adapt to unseen datasets, using patterns learned from multiple domains. Unlike traditional models that need retraining for every new dataset, HeLNMs generalize relationships across different sources, allowing them to infer outcomes for new data without additional computational overhead. This capability ensures real-time adaptability, making it suitable for rapidly evolving industrial applications.

One of the key advantages of HeLNMs is their ability to make accurate predictions even when the dataset is sparse or small. This is achieved through cross-domain knowledge transfer, where insights from related datasets compensate for missing data. Additionally, synthetic data generation using CTGAN and VAE fills data gaps, maintaining prediction reliability. SXI++ further enhances accuracy by identifying strong correlations across features, making it effective for both regression and classification tasks, even with limited training samples. These techniques enable HeLNMs to function effectively without large, labelled datasets.

3.2.2.2 Role of CTGAN and VAE in HeLNMs

Two key deep learning architectures, Conditional Generative Adversarial Networks (CTGANs) and Variational Autoencoders (VAEs), play crucial roles in HeLNMs:

- Conditional Generative Adversarial Networks (CTGANs)—CTGANs are used to generate synthetic data samples to fill in gaps within the dataset, ensuring that HeLNMs have a balanced and well-represented data distribution.

- Variational Autoencoders (VAEs)—VAEs aid in feature extraction and data imputation, particularly in cases where missing values occur due to inconsistent reporting across different domains.

By incorporating these techniques, HeLNMs improve data completeness and model robustness, enabling more reliable predictions and insights.

3.2.2.3 Enhancing HeLNMs with CTGAN and VAE for Data Augmentation and Completion

Conditional Generative Adversarial Networks (CTGAN) plays a crucial role in enriching HeLNMs by generating synthetic data that closely mirrors the original dataset. Since HeLNMs integrate data from multiple domains, there can be gaps in feature representation due to missing values or imbalanced distributions. CTGAN leverages a conditional generative adversarial network (GAN) framework to produce high-quality synthetic samples that maintain the statistical relationships of categorical and numerical features. By conditioning the generator on specific attributes, CTGAN ensures that the generated data preserves critical dependencies, allowing HeLNMs to retain domain-specific information even in the absence of complete real-world data.

Beyond filling gaps, CTGAN enhances HeLNMs by addressing class imbalance issues that commonly arise when merging heterogeneous datasets. In scenarios where certain feature combinations are underrepresented, CTGAN generates synthetic samples to balance the distribution, preventing biases in downstream machine learning models. Parameters such as gen_x_times and pregeneration_frac allow for controlled augmentation, ensuring that the synthetic data does not overwhelm the real data while providing sufficient diversity to improve model generalization. Additionally, feature-matching loss ensures that generated data aligns closely with the distribution of real data, making CTGAN-enhanced datasets more reliable for predictive modelling in HeLNMs.

Variational Autoencoders (VAE), on the other hand, focuses on data completion by imputing missing values in HeLNMs through its ability to learn latent representations of the dataset. By compressing data into a lower-dimensional latent space and then reconstructing it, VAE captures underlying patterns in complex datasets, allowing it to generate plausible estimates for missing entries. This is particularly valuable for HeLNMs, where datasets from different domains may have non-overlapping records due to data sparsity. Unlike simple imputation methods that rely on averages or nearest neighbours, VAE-generated values maintain consistency with the original dataset's structure, ensuring that HeLNMs retain meaningful insights rather than introducing statistical noise.

Moreover, VAE's reconstruction process is enhanced by its KL divergence loss, which ensures that the latent space follows a structured probability distribution. This makes the generated missing values more reliable, as they are not just interpolations but learned representations of underlying trends. The ability to restore data to its original scale further ensures that the imputed values maintain real-world relevance. By leveraging VAE's powerful feature extraction and reconstruction capabilities, HeLNMs can achieve a completer and more robust dataset, reducing uncertainties caused by missing data and improving predictive accuracy.

Together, CTGAN and VAE complement each other in HeLNMs by tackling both synthetic data generation and missing value imputation. CTGAN provides a diverse and balanced dataset by creating realistic synthetic samples, while VAE ensures dataset integrity by filling missing values with high-fidelity estimates. This combination enables HeLNMs to extract deeper and more actionable insights by mitigating data sparsity and bias, making them an invaluable tool for industries where external factors play a crucial role in decision-making. Ultimately, the synergy of CTGAN and VAE strengthens the HeLNM framework, allowing for more accurate, insightful, and comprehensive analysis across diverse domains.

3.2.2.4 Finding Target Feature Missing Values Using SXI++ Correlation or Voting Classifier

Heterogeneous Large Numerical Models (HeLNMs) enable the prediction of missing target variables in datasets where no reference dataset with actual target values is available for validation. This is achieved by leveraging either a Voting Classifier for classification tasks or SXI++ correlation-based predictions, which can be applied to both classification and regression scenarios. The process begins by training a model using the HeLNM dataset, which contains the target variable along with a set of common features that overlap with the test dataset. For classification, the Voting Classifier integrates multiple base models, such as XGBoost, Random Forest, SVM, KNN, and Logistic Regression, using a soft voting mechanism to determine the most probable class. For regression or mixed scenarios, SXI++ utilizes correlation-based analysis to infer missing values by identifying patterns within the available features.

Since no reference dataset with actual target values exists for accuracy validation, an internal verification approach is applied. The predicted target values are used to train a secondary model within the HeLNM framework, which is then tested against different subsets of the data to evaluate consistency and stability. This iterative approach ensures that the generated target variable aligns with the underlying data patterns. By combining ensemble learning techniques and SXI++ correlation analysis, HeLNMs provide a scalable solution for handling missing target variables, making them highly effective for real-world applications where complete datasets are often unavailable.

3.2.2.5 Examples of Actionable Insights from HeLNMs and Their Comparison with HoLNMs

Given that HeLNMs incorporate a wider range of factors, they enable the extraction of actionable insights that go beyond what homogeneous LNMs can achieve. Some key comparative analyses include:

1. Impact of External Factors on Cement Production—Unlike homogeneous LNMs, HeLNMs consider fluctuations in electricity generation costs, variations in raw material availability, and geopolitical influences on supply chains.

2. Optimization of Alternative Cement Formulations—The model evaluates how different combinations of SHS, Fly Ash, and Slag perform under varying conditions, leading to optimized formulations that balance cost, performance, and sustainability.
3. Energy Efficiency and Carbon Footprint Analysis—By integrating multiple energy sources, HeLNMs provide a detailed analysis of carbon emissions and energy consumption, allowing for the development of eco-friendly cement production strategies.
4. Predictive Maintenance and Risk Mitigation—HeLNMs identify potential supply chain disruptions or operational inefficiencies, enabling proactive decision-making.

The automated decision support system within HeLNMs-SXI++ translates insights into direct operational improvements. Decision trees and optimization models suggest key changes in raw material composition, recommend energy-efficient techniques, and provide scenario-based simulations to test various supply chain strategies. These measures lead to tangible performance improvements, ranging from 20% to 90%, depending on the implementation scale and industry conditions.

Through HeLNMs-SXI++, manufacturers gain a competitive advantage by dynamically optimizing their processes. The system's ability to analyse heterogeneous factors, predict outcomes, and drive real-time decisions ensures enhanced production efficiency, sustainability, and cost-effectiveness. This advanced approach revolutionizes manufacturing, positioning companies for long-term success in an evolving market.

3.2.2.6 Major Hurdles for HeLNMs Commercialization

To achieve successful development and commercialization of HeLNMs, several challenges must be addressed:

1. **Scalability of HeLNMs**—Ensuring the model architecture efficiently handles large, heterogeneous datasets with varying feature distributions, missing values, and inconsistencies.
2. **Robust Target Variable Imputation**—Improving SXI++ and Voting Classifier methods to infer missing target variables with high confidence, particularly when no reference dataset exists.
3. **Cross-Domain Generalization**—Developing adaptive learning mechanisms that allow HeLNMs to perform well across different domains without significant retraining.
4. **Computational Efficiency**—Optimizing model execution time for real-time or near-real-time decision-making while handling high-dimensional datasets.
5. **Validation Without Ground Truth**—Establishing alternative validation methods when actual target values are unavailable, ensuring reliability in model predictions.
6. **Integration with Decision Support Systems**—Enhancing interpretability by integrating explainability techniques to ensure model decisions align with human intuition.

Digital Financial Fraud

4

4.1 Introduction

Explosive growth in digital financial services and online banking has transformed the way customers access credit, open accounts, and conduct transactions—providing speed, convenience, and scale. But the digital revolution has also resulted in explosive growth of fraudulent schemes, primarily in the guise of spurious or doctored bank account openings. These often use systemic weaknesses through the use of stolen identities, synthetic identities, or misrepresentation, to obtain unauthorized access to financial services for illegal monetary benefits. These schemes not only cause billions of dollars in loss every year, estimated at over $50 billion worldwide—they also undermine consumer confidence and test the efficacy of the financial system.

In mass market digital onboarding environments, particularly in consumer and fintech banks, the bad actors are increasingly bypassing the legacy verification procedures. These attacks render inevitable the implementation of next-generation, scalable, and intelligent fraud detection solutions that can handle sophisticated, fast-changing attack patterns.

Conventional detection of fraud—using rule-filter-based or static machine learning methods—frequently fails to identify sophisticated schemes of fraud. These technologies are impaired by the restrictive scope of features, the high percentage of false positives, and the inability to adapt easily to new attack methods. Although present-day AI and ML technologies hold much potential, numerous existing implementations are missing in interpretability, the ability to generalize across domains, and the ability to learn dynamically, rendering them unsuitable for real-world applications in the face of heterogenous financial settings. To surmount the aforementioned shortcomings, the present work puts forth the Large Numerical Model (LNM), powered by the proprietary Sriya Expert Index (SXI) scoring system, constructed by Sriya.AI, LLC. SXI integrates the results of 5–10 machine

S. Kilambi, M. Banavar, *Large Numerical Models from a Business Perspective*, Synthesis Lectures on Technology Management & Entrepreneurship,
https://doi.org/10.1007/978-3-032-14869-8_4

learning algorithms within a single, real-time risk of fraud score, employing behavioral, transactional, demographic, and temporal signals. Six diverse fraud datasets—Bank Account, Credit Card, IEEE-CIS, Simulated Fraud, Mobile Transaction, and Cryptocurrency Transaction fraud—were integrated into a common Master Dataset of 509,841 records with 394 features. Missing values were handled using Variational Autoencoder (VAE) based imputation to preserve distributional integrity.

Master Train was employed to train the LNM, while AutoML-driven XGBoost was designated the baseline for comparative analysis. It enables the model, by multi-source, high-dimensional training, to adapt well to different kinds of frauds, as well as different platforms.

These results confirm that SXI greatly outperforms baseline models—approaching near-perfection in accuracy, precision, and AUC values, even under class imbalance—suggesting that data-driven, real-time scoring of suspected fraud rates can be highly effective in reducing monetary risk. Suspected fraud rate reductions of 20% (short term), over 50% (mid-term), and 90% (long term) support this observation. Decision tree visualizations also produce actionable advice, pointing out the chief identifiers of fraud such as document discrepancies, poor credit signals, and behavioral anomalies.

By combining multiple sources of fraud, real-time scoring, deep learning imputation, and explainable schemes of logic, the SXI-powered LNM offers a next-generation solution to the detection of fraud that is robust, clear, and responsive to evolving financial environments.

4.2 Materials and Methods

4.2.1 Data Description

The dataset used for this fraud detection analysis was constructed by merging six diverse fraud datasets into a unified Master Dataset as part of a comprehensive approach to building a Large Numerical Model (LNM). These datasets include: Bank Account Fraud, Credit Card Fraud, IEEE-CIS Fraud, Simulated Credit Card Fraud, Mobile Transaction Fraud, and Cryptocurrency Transaction Fraud, collectively comprising 509,841 records and 394 features. This combined dataset was designed to capture a wide range of fraud patterns across different financial domains, reflecting realistic and complex transaction behaviors.

To develop and evaluate the LNM effectively, the dataset was partitioned into two distinct subsets (see Table 4.1) for training and testing.

Table 4.1 Fraud dataset distribution across training and testing sets

Fraud status	Training	Testing	Total
Total	284,522 (70%)	123,350 (30%)	407,872 (100%)

To address missing data while preserving the underlying feature relationships, Variational Autoencoder (VAE)-based imputation was applied across all datasets, maintaining consistency in distribution and enhancing data quality for both training and evaluation. The merged dataset included a balanced mix of numerical and categorical features, with key attributes such as transaction amount, timestamp, user demographics, digital footprint indicators, and temporal patterns. Three sensitive features—age group, employment status, and debt-to-income ratio—were retained for fairness evaluation in machine learning.

By integrating multi-source fraud data and applying advanced preprocessing and training strategies, the LNM framework provides a highly scalable and generalizable solution for real-time, cross-domain fraud detection, capable of learning from heterogeneous patterns while maintaining high precision and interpretability.

4.2.2 Descriptive Statistics

4.2.2.1 Fraud Detection Status Visualization

The below pie chart (Fig. 4.1) shows two segments: "Not Fraud" and "Fraud." The *Not Fraud* category dominates the chart, represented by a teal-colored segment accounting for 96.93% of the total, indicating the overwhelming proportion of transactions identified as legitimate. In contrast, the *Fraud* category, shown in a yellow segment, makes up only

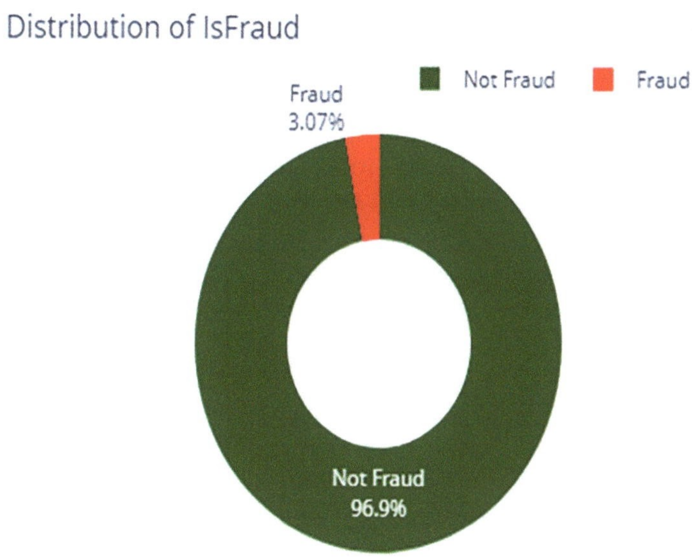

Fig. 4.1 Fraud detection status. This donut chart displays the distribution of fraud detection outcomes, with 96.9% of applications classified as "Not Fraud" (green) and 3.07% as "Fraud" (Red), reflecting the SXI model's enhanced detection rate

3.07%, highlighting the small but critical portion of cases identified as fraudulent. This visual representation emphasizes the highly imbalanced nature of the dataset, underscoring the challenge faced by the fraud detection model in accurately identifying the minority class. Despite the imbalance, the model must maintain high precision and recall to ensure even the rare fraudulent instances are effectively detected.

4.2.3 Method

SXI Score Calculation, Correlation with Target Feature, Using SXI/SXI++ to predict future outcomes and finally the usage of SXI/SXI++ to improve business outcomes has been described in Chap. 3, SXI/SXI++.

SXI/SXI++ "Score, Correlate, Predict, Improve" process overview below:

SXI stands for moving score/index that is based on a proprietary formula with weights for 5–10 ML algorithms. SXI stands for super feature and it is a weighted true representational version of all key features that convert a multi-dimensional hard issue to a simpler 2-dimensional solution. SXI with proprietary deep neural network algorithm comprises an iterative process where continuously algorithm weights are updated considering maximum weights given by the 5–10 ML algorithms. This process tries to refine the correlation of SXI scores to business results so that accuracy and delineation are enhanced.

4.3 Results and Discussions

From the figure (Fig. 4.2), a key finding was the discriminative power of the SXI score, which revealed a clear separation boundary at SXI = 0.68:

- All transactions with SXI > 0.68 were 100% non-fraudulent
- All fraud cases occurred when SXI ≤ 0.68

This sharp boundary makes the SXI score a powerful linear classifier for binary fraud prediction. Notably, this level of delineation was achieved without overfitting, as verified by cross-validation.

As illustrated in (Fig. 4.3), the correlation coefficient of 0.91 (negative) between SXI scores and the Fraud label (isFraud) indicates an exceptionally strong inverse relationship between the two variables. In practical terms, this implies that as the SXI score increases, the likelihood of a transaction being fraudulent decreases proportionally. Conversely, lower SXI scores are strongly associated with a higher probability of fraud. Therefore, any upward movement in SXI directly contributes to reduced fraud exposure.

This high negative correlation serves as clear evidence of SXI's predictive power. In contrast to traditional fraud indicators that often require complex feature engineering, SXI

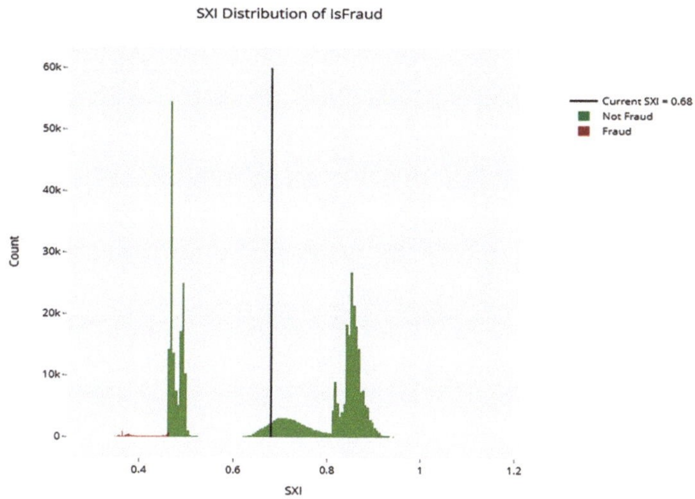

Fig. 4.2 SXI delineation

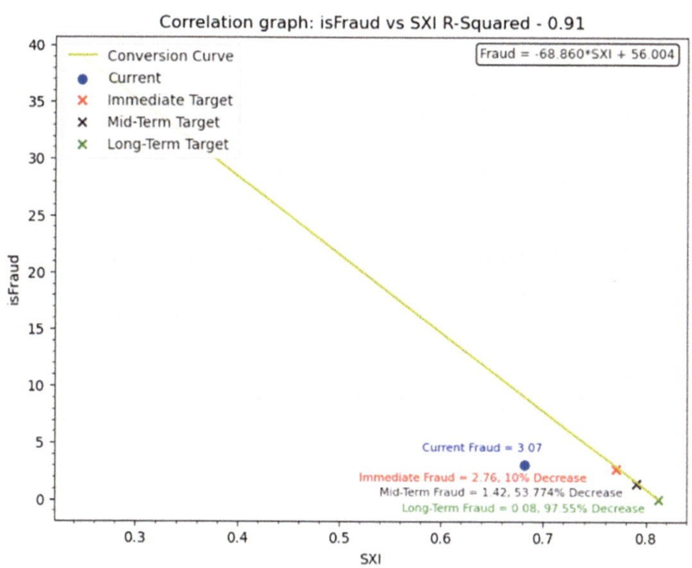

Fig. 4.3 Correlation plot

alone encapsulates much of the behavioral and transactional variance that contributes to fraud detection.

The relationship between SXI score improvements and fraud reduction is not only statistically significant but also business actionable. Based on regression modeling and outcome mapping, we observe the following pattern of improvement (Table 4.2):

Table 4.2 SXI vs Projected fraud rate: Recommendations for improvement

SXI increase	Projected fraud rate	Relative fraud reduction
Current (0.68)	3.07%	–
↑ by 11.76% (0.77)	2.76%	↓ 10.1%
↑ by 14.71% (0.79)	1.42%	↓ 53.77%
↑ by 20.59% (0.81)	0.08%	↓ 97.55%

- A modest increase in SXI scores can already yield meaningful reductions in fraud, especially useful for short-term, tactical interventions (e.g., adjusting customer verification steps or rule filters).
- Mid-term improvements yield transformational fraud control, cutting losses by over 50%.
- Long-term improvements could practically eliminate fraud, achieving near-zero false approvals in high-risk segments.

The **current decision tree** (Fig. 4.4) identifies key patterns behind the existing fraud rate. Transactions are more likely to be classified as *Not Fraud* when users reside in better-quality housing (ELEVATORS_MEDI > 0.07), have stable employment histories (DAYS_EMPLOYED > 135.5), and live in standard apartments with minimal non-residential units (NONLIVINGAPARTMENTS_MEDI ≤ 0.001). In contrast, *Fraud* cases are associated with low infrastructure areas, smaller transaction values (AMT_GOODS_PRICE ≤ 677,250), and poor credit indicators (EXT_SOURCE_3 ≤ 0.536).

The **target decision tree** (Fig. 4.5) outlines paths to reduce fraud by **10%**, lowering it to **2.76%**. It shows that *Not Fraud* classifications increase when applicants have high external trust scores (EXT_SOURCE_1 > 0.69 and especially >0.98) and simple housing (LIVINGAREA_MODE ≤ 0.04). Meanwhile, *Fraud* is more likely when users show weak credit signals (EXT_SOURCE_1 ≤ 0.69), document inconsistencies (FLAG_DOCUMENT_3 > 0.6), or low external reliability (EXT_SOURCE_3 ≤ 0.67). These refined conditions guide targeted interventions to reduce fraud effectively.

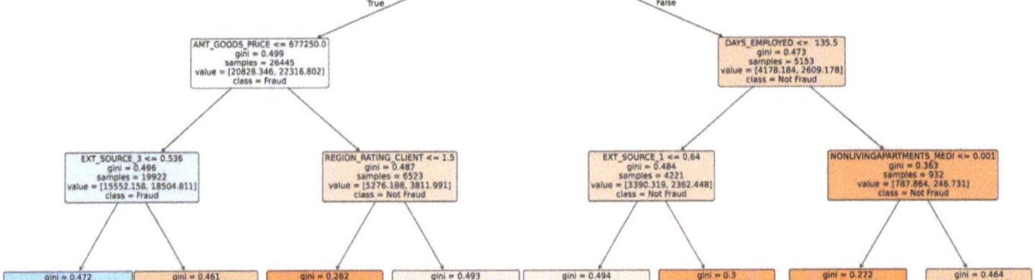

Fig. 4.4 Current decision tree

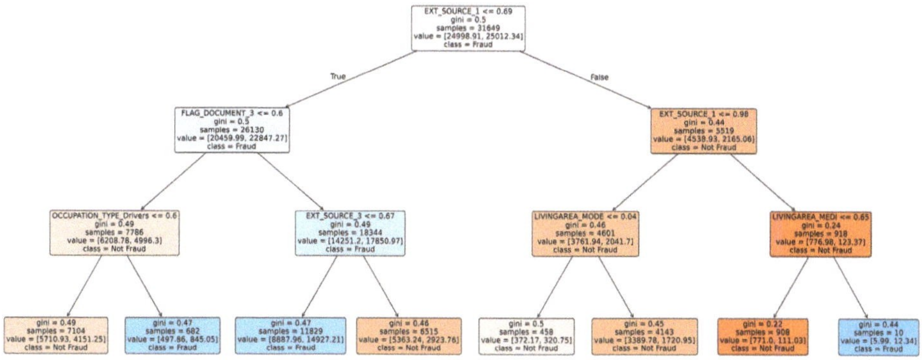

Fig. 4.5 Target decision tree

Table 4.3 Performance metrics table

Metrics	Credit card	Banking + Credit card	Banking + Credit + Mobile	Banking + Credit + Mobile + IEEE	Banking + Credit + Mobile + IEEE + Simulated
Accuracy	99.58%	96.4%	98.03%	97.6%	99.2%
Precision	100%	100%	97.03%	98.1%	100%
AUC	0.96	0.83	0.88	0.91	0.95
Recall	90%	62.9%	70.9%	78.3%	87.2%

The model's performance improves as additional data sources are integrated, with the best results seen in the most enriched dataset—Banking + Credit + Mobile + IEEE + Simulated (Table 4.3).

- Accuracy reflects overall correctness. It remains consistently high across all datasets, peaking at 99.58% for credit card data and 99.2% with simulated features, showing strong general performance.
- Precision indicates how many of the flagged fraud cases were actually fraud. It stays nearly perfect (100%) in most setups, meaning the model rarely makes false fraud accusations.
- AUC (Area Under Curve) measures the model's ability to distinguish between fraud and non-fraud. It improves with data complexity, from 0.83 (basic banking + credit) to 0.95 (full feature set), showing better fraud separation.
- Recall represents how many actual fraud cases were correctly detected. While it starts lower (62.9%) with limited data, it climbs to 87.2% with simulated features, highlighting the model's growing sensitivity to true fraud.

4.4 Conclusions

Unlike traditional machine learning models that are trained once and applied uniformly, the SXI scoring system dynamically evaluates each transaction based on its individual characteristics. This individualized scoring approach captures the nuanced risk factors associated with fraudulent behavior. The strong inverse correlation between SXI scores and fraud outcomes enables the use of SXI as a reliable signal for real-time fraud detection and strategic thresholding. These insights guide decision tree structures aimed at achieving targeted fraud reduction goals, supporting tactical, mid-term, and long-term business interventions. As new transactions are processed, SXI scores allow institutions to determine, with high precision, the likelihood of fraud before decisions are finalized.

The results demonstrate the remarkable fraud detection capabilities of the Sriya Expert Index (SXI) and its deep learning-enhanced version, SXI++. Using a comprehensive dataset of over 400,000 transactions, the model achieved outstanding predictive performance—including up to 99.58% accuracy, 100% precision, and an AUC score of 0.96–0.95 across multiple datasets. These results significantly outperform traditional machine learning models such as XGBoost, Random Forest, and hybrid approaches like Mixture of Experts, showcasing the superiority of the SXI++ framework for fraud detection.

The study also introduces a structured, measurable strategy for reducing fraud rates based on SXI thresholds. Starting from a baseline fraud rate of 3.07%, short-term improvements in SXI scores yield a 10% reduction, with mid-term and long-term enhancements achieving reductions of over 50% and up to 97.5%, respectively. These outcomes reflect the compounding benefits of sustained score optimization, reinforcing SXI's role in not only detecting but actively mitigating fraud over time.

Despite these strong results, certain limitations are recognized. The current analysis is based on transactions from a specific credit environment, which may limit generalizability to other financial ecosystems or fraud contexts. Additionally, external influences such as market volatility, device spoofing techniques, or emerging fraud patterns were not explicitly modeled in this framework.

Future research should focus on expanding the SXI framework's applicability by incorporating diverse datasets from other institutions, payment channels, or geographies. Introducing behavioral signals from real-time user interactions, digital footprints, or device telemetry could further enrich fraud detection capabilities. Finally, the integration of reject inference—examining transactions flagged or declined—would help uncover patterns in attempted fraud and improve the model's defensive coverage against evolving threats.

In conclusion, SXI++ represents a significant advancement in intelligent fraud prevention, offering a scalable, real-time, and explainable framework that empowers institutions to take informed action, minimize losses, and build resilient, data-driven defenses.

5 Reducing Late Payments in Peer-to-Peer Lending

5.1 Background and Introduction to Data Source "Lending Club"

Traditional financial institutions have high thresholds, low returns, and long investment cycles, driving investors to seek more flexible, higher-yield, and lower-amount investment options. However, limitations in the personal credit information system and slow credit infrastructure development prevent some deserving individuals from accessing funds, leading to the rise of streamlined platforms like P2P lending. P2P platforms act as intermediaries, connecting borrowers and lenders online. They offer convenient and accessible lending and money transfer services. Additionally, deep learning techniques provide valuable insights for interpreting various forms of data, potentially enhancing the functionality of P2P platforms [1]. The demand from both investors and borrowers has fueled the growth of P2P lending platforms significantly, contributing to their rapid expansion. These platforms have become integral to the financial market in my country, fulfilling crucial roles in meeting investment and financing demands while fostering market diversity. However, despite their importance, P2P platforms suffer from deficiencies in financial management and risk control capabilities, leading to a considerable number of defaulted loans [2].

The largest P2P lending platform in the United States, known as "Lending Club" operates as an online platform that bypasses traditional financial institutions, linking numerous investors with potential borrowers to facilitate capital investment and credit borrowing. Borrowers can submit loan requests detailing the purpose of the loan along with their personal financial details. Investors have the flexibility to select the amount of capital they wish to invest and can also choose the specific borrower.

Previous studies predominantly employed machine learning techniques to address the challenge of predicting late payments in peer-to-peer (P2P) lending platforms, focusing primarily on identifying high-risk borrowers. However, these approaches often lacked

S. Kilambi, M. Banavar, *Large Numerical Models from a Business Perspective*, Synthesis Lectures on Technology Management & Entrepreneurship, https://doi.org/10.1007/978-3-032-14869-8_5

interpretability regarding the influence of key borrower characteristics and financial behaviors on late payment predictions, failed to propose effective strategies for reducing late payments over short-, mid-, and long-term periods, and relied on models that required extensive prior training. Additionally, there was limited integration of advanced scoring systems, which could leverage diverse borrower data, such as credit history, income stability, and repayment trends, to more accurately assess late payment risks across various lending scenarios. By incorporating a comprehensive scoring system, the potential for more precise and actionable risk analysis is greatly enhanced, offering lenders a holistic understanding of borrower repayment behavior.

The inability of borrowers to repay loans poses a significant challenge to the sustainability of the peer-to-peer (P2P) lending sector. This study leverages predictive modeling techniques to analyze historical applicant data from Lending Club, focusing on reducing late payments through the proprietary Sriya Expert Index (SXI) AI-ML algorithm. SXI serves as a super feature, synthesizing the outputs of 5–10 machine learning algorithms into a simplified score/index, enabling accurate prediction of late payments. The model dynamically adjusts algorithmic weights to optimize precision, considering critical features such as credit history, income, and repayment behavior. In comparison to traditional machine learning models, the Sriya Expert Index (SXI) algorithm significantly outperforms established approaches in predicting late payments. Models such as Random Forest and XGBoost achieved accuracies of 78.80% and 84.52%, respectively, while the Mixture of Experts (MOE) neural network reached 92.10% [3]. The Support Vector Machine (SVM) with a linear kernel delivered an AUC of 0.935, slightly higher than the 0.92 AUC of XGBoost. However, SXI surpasses all these models with a near flawless accuracy of 99.80% and an AUC score of 0.998. This demonstrates the model's superiority in identifying late payment risks and its potential to guide effective intervention strategies. One of the standout features of SXI enabled AI-ML is to improve business outcomes with actionable insights. The study outlines a phased methodology for reducing late payments (desired business outcome), achieving an initial 20% reduction and further improvements to 50% and 80% in the mid-term and long-term, respectively. These findings highlight the transformative potential of SXI in enhancing risk management in P2P lending, offering a scalable, data-driven solution to improve the financial health of the sector.

To address these gaps, the study in this chapter focuses on: (1) evaluating the performance of the proprietary SXI algorithm developed by Sriya.AI, LLC as a multivariate scoring system to predict late payment risk as a binary classification problem, (2) enhancing the SXI scoring methodology using a Proprietary Deep Neural Network algorithm and correlating it with late payment rates over a defined period, and (3) employing a targeted decision tree framework to interpret the most effective pathways leading to both high- and low-risk borrowers, thereby offering evidence-based recommendations for reducing late payment incidences and improving the financial stability of P2P lending platforms.

5.2 Materials and Methods

5.2.1 Data Description

The source of the data for this paper on predicting and reducing the impact of late loan payments using Lending Club's historical loan applicant information likely comes from Lending Club's internal database or archives dated from 2007 to 2020 Q3. Lending Club is a US peer-to-peer lending company, headquartered in San Francisco, California. It was the first peer-to-peer lender to register its offerings as securities with the Securities and Exchange Commission (SEC), and to offer loan trading on a secondary market.

The data given contains information about past loan applicants and whether they have late or not yet defaulted on their payments. The aim is to identify patterns which indicate if a person is likely to make late payments, which may be used for takin actions such as denying the loan, reducing the amount of loan, lending (high risky applicants) at a higher interest rate, etc.

The initial dataset consists of 2,260,668 rows and 151 features and through further data cleaning 58 features had null values exceeding 35% of the total entries. Given the significant proportion of missing data in these features was dropped from the dataset which resulted in a reduction in the number of features to 93 which will be used for further data analysis.

The target variable for this analysis is the Loan Status (Fig. 5.1), which has two categories: "Current" and "Late Payments" from the reformatted data. The "Current" category represents applicants who are in the process of paying their loan installments, meaning the loan tenure has not yet been completed. The "Late-Payments" category represents applicants who have paid their installments after the due date. To ensure a balanced dataset for model building, a random sample of 26,000 records was taken from the "Current" category and combined with the 25,816 records from the "Late Payments" category (Late (31–120 days) and Late (16–30 days)), resulting in a final dataset of 51,816 rows.

```
Out[104]: loan_status
          Fully Paid                                                 1076751
          Current                                                     878317
          Charged Off                                                 268559
          Late (31-120 days)                                           21467
          In Grace Period                                               8436
          Late (16-30 days)                                             4349
          Does not meet the credit policy. Status:Fully Paid            1988
          Does not meet the credit policy. Status:Charged Off            761
          Default                                                         40
          Name: count, dtype: int64
```

Fig. 5.1 Loan status. We consider three categories, Current, Late (31–120 days), and Late (16–30 days)) to represent the loan status

The dataset contains a mix of categorical and numerical features. There are 16 categorical features, and 69 numerical features. It's worth noting that two features, both categorical and numerical, had only one unique value across the entire dataset, which may indicate limited variability in those features. In addition to the categorical and numerical features, the dataset also includes eight features that are related to dates, IDs, zip codes, and other non-predictive information. The categorical features are one hot encoded thus increasing the number of features from 82 to 171 features. The final dataset used for model building will consist of 51,816 rows and 171 features, after excluding 35% of missing values and the non-predictive features.

5.2.2 Descriptive Statistics

5.2.2.1 Missing Values

The dataset reveals significant missing values in the top ten features (Fig. 5.2), with *open_acc_6m* having the highest at 85.36%, followed by *open_il_12m* (84.11%). Key secondary applicant features like *FICO range (high)* (79%) and *FICO range (low)* (78.01%) also exhibit substantial missingness. Other notable features, including *open_il_24m* (78.12%), *open_act_il* (73.97%), and *sec_app_inq_last_6mths* (70.25%), show similar trends. Features such as *sec_app_open_acc* and *sec_app_revol_util* (both 68.04%) and *sec_app_open_act_il* (66.66%) indicate critical gaps in credit-related attributes. Overall, 58 features with missing values exceeding 35% of total entries were dropped from the dataset. This reduction resulted in 93 features being retained for further data analysis, ensuring better data quality and improved modeling potential.

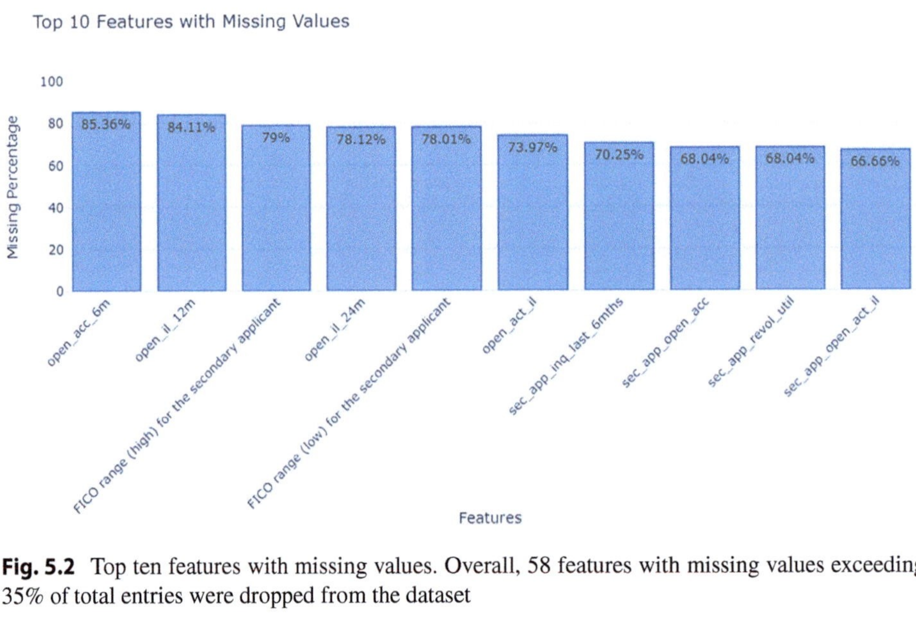

Fig. 5.2 Top ten features with missing values. Overall, 58 features with missing values exceeding 35% of total entries were dropped from the dataset

5.2.2.2 Loan Amount vs Interest Rate vs Loan Status

The descriptive statistics for loan amounts indicate a noticeable difference in the distribution between Current and Late payments. On average, borrowers with late payments tend to take slightly higher loan amounts ($17,021.50) compared to those who are current ($15,926.56). Additionally, (Fig. 5.3) the interquartile range (IQR) is broader for late payments ($14,000: from $10,000 to $24,000) than for current payments ($13,125: from $8400 to $21,525), suggesting that late payments are associated with more variability in loan sizes. This difference may point to a pattern where larger loans (Fig. 5.4) or a certain borrower profile could be contributing to increased likelihood of delinquency.

The correlation matrix (Fig. 5.5) highlights a moderate positive relationship between interest rates and loan status (correlation = 0.27). This suggests that borrowers with higher interest rates are more likely to have late payments. However, the correlation between loan amounts and loan status (0.056) is weaker, indicating that loan size alone is not a strong predictor of payment timeliness. The weak positive correlation between interest rates and loan amounts (0.08) suggests that while larger loans may marginally attract higher rates, this is not a dominant trend. Taken together, the insights imply that interest rates might serve as a more reliable indicator of payment risk than loan amounts, pointing to potential areas for targeted risk mitigation.

5.2.2.3 Loan Status Distribution

The dataset used for the analysis is well-balanced across the different splits, ensuring reliable and unbiased model training, testing, and validation (Table 5.1). The training set comprises 36,271 records, nearly evenly divided between Late Payments (18,146 or 50.03%) and Current Payments (18,125 or 49.97%). This balance is critical for developing

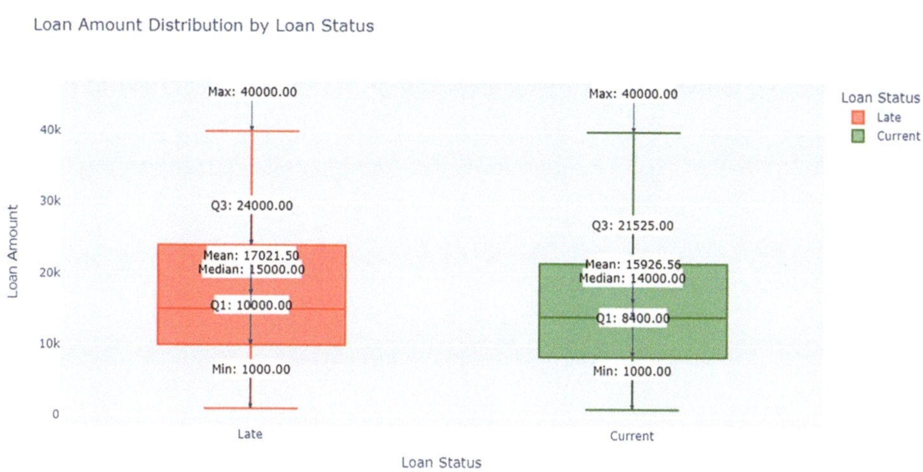

Fig. 5.3 Loan amount distribution w.r.t loan status. This boxplot displays the distribution of loan amounts by loan status, highlighting key statistics such as the median, Q1, Q3, and extremes to compare current and late payments

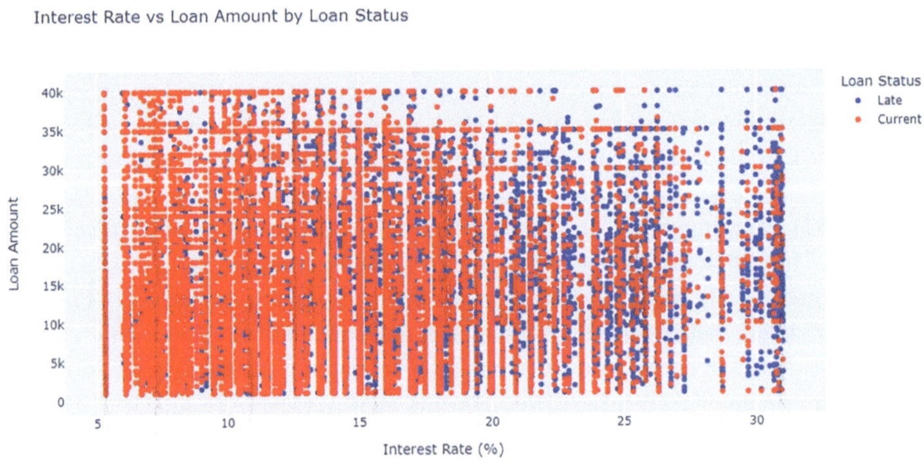

Fig. 5.4 Interest rate vs loan amount w.r.t loan status. This figure visualizes the relationship between interest rates and loan amounts, highlighting differences in trends for borrowers classified as current versus late payments

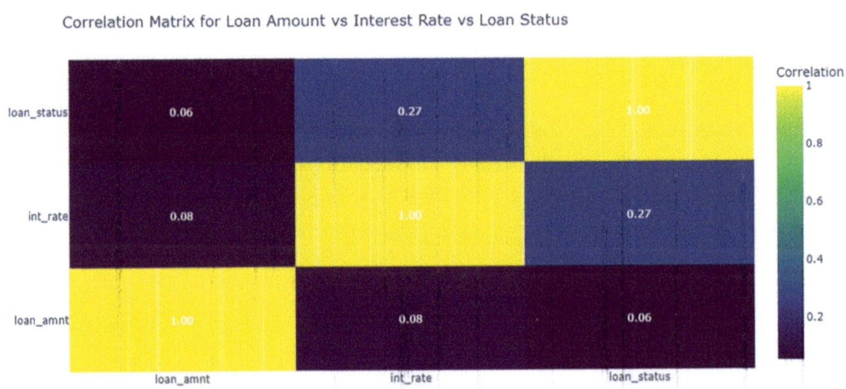

Fig. 5.5 Correlation of loan amount and interest rate wr.t loan status. This figure examines the correlation between loan amounts and interest rates, providing insights into how these variables interact for current and late payment borrowers

Table 5.1 Loan status distribution

Loan status	Training	Testing	Validation	Total
Late payments	18,146 (50.03%)	5088 (49.10%)	2582 (49.83%)	25,816 (49.82%)
Current payments	18,125 (49.97%)	5275 (50.90%)	2600 (50.17%)	26,000 (50.18%)
Total	36,271	10,363	5182	51,816

a model that does not favor one class over the other, particularly for binary classification problems like loan payment prediction.

Similarly, the testing set of 10,363 records maintains a near-even distribution, with 49.10% (5088 records) classified as Late Payments and 50.90% (5275 records) as Current Payments. The validation set also follows this trend, with 49.83% (2582 records) representing Late Payments and 50.17% (2600 records) as Current Payments. This consistent class distribution across all subsets (training, testing, and validation) ensures that the model's performance metrics, such as accuracy and precision, are not influenced by an imbalanced dataset, thereby enhancing the credibility of the results. Overall, the dataset's balance supports robust statistical analysis and reliable model evaluation.

5.2.3 Method

SXI Score Calculation, Correlation with Target Feature, Use of SXI/SXI++ to predict future outcomes and finally the use of SXI/SXI++ to improve business outcomes has been detailed in the SXI/SXI++ Chap. 3. The summary of SXI/SXI++ "Score, Correlate, Predict, Improve" methodology is as follows:

SXI is a dynamic score/index obtained from a proprietary formula consisting of weights from 5–10 ML algorithms [4]. SXI is a super feature and is a true weighted representative of all important features which converts a multi-dimensional hard-to-solve problem into a simpler 2-dimensional solution. SXI with Proprietary Deep Neural Network algorithm involves an iterative approach where it dynamically adjusts algorithm weights based on the most significant weights provided by the 5–10 ML algorithms. This process aims to enhance the correlation between SXI scores and business outcomes, thereby improving accuracy and delineation.

5.3 Results and Discussions

The SXI stands at 0.97, serving as the benchmark score. In terms of overall payments, 49.82% are categorized as good outcomes (current payments), while 50.18% are identified as bad outcomes (late payments). According to the initial delineation based on the benchmark, there is a slight inclination towards categorizing late payments as above SXI and current payments as below SXI (Fig. 5.6).

The SXI algorithm demonstrated outstanding performance in classifying loan payment statuses, achieving excellent metrics across both the test and validation sets. In the test set (Fig. 5.7), with 10,363 records, the model correctly classified 5222 current payments and 5037 late payments. Only 53 current payments were misclassified as late payments, and 51 late payments were misclassified as current payments. This resulted in an impressive accuracy of 99.80% and an AUC of 0.998 (Fig 5.8), reflecting the model's accuracy and precision in predicting loan payment outcomes.

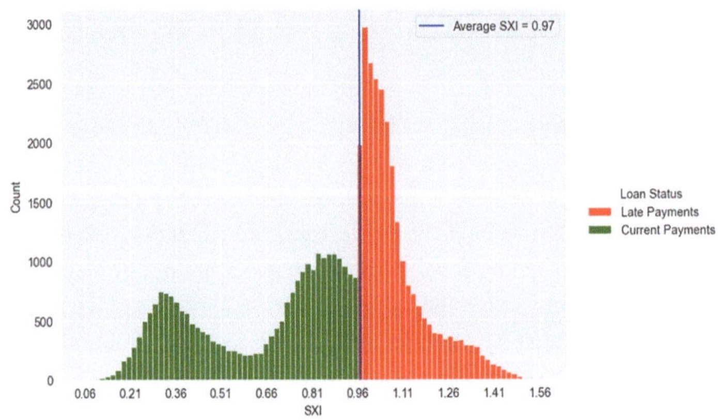

Fig. 5.6 SXI distribution. This figure shows the distribution of SXI scores, illustrating how scores differ between borrowers likely to make timely repayments and those at risk of late payments

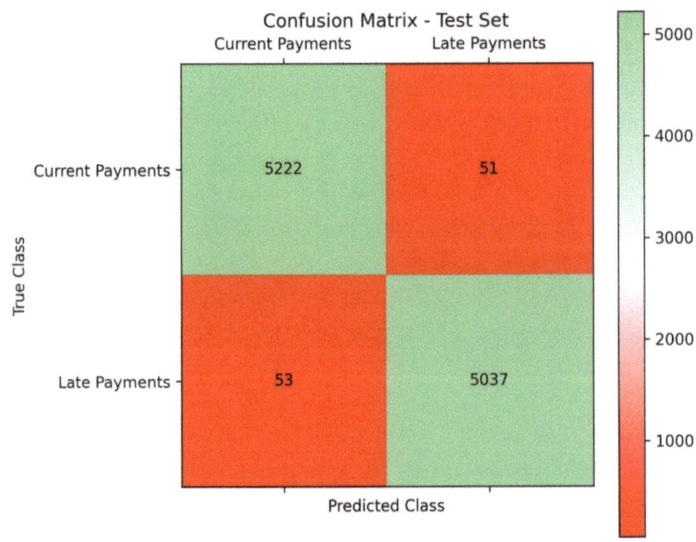

Fig. 5.7 Confusion matrix—test set

Similarly, in the validation set of 5182 records, the SXI algorithm maintained its exceptional performance. It correctly classified 2594 current payments and 2578 late payments, with only six current payments misclassified as late payments and four late payments misclassified as current payments (Fig. 5.9). This also resulted in an accuracy of 99.80% and an AUC of 0.998 (Fig. 5.10), demonstrating the robustness of the SXI algorithm. These results underscore its effectiveness in accurately predicting loan payment statuses, establishing the SXI algorithm as a reliable tool for applications where precise classification is crucial.

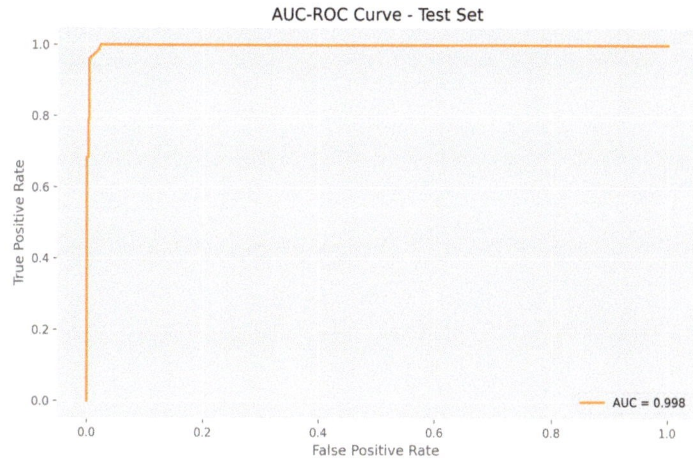

Fig. 5.8 AUC-ROC graph—test set

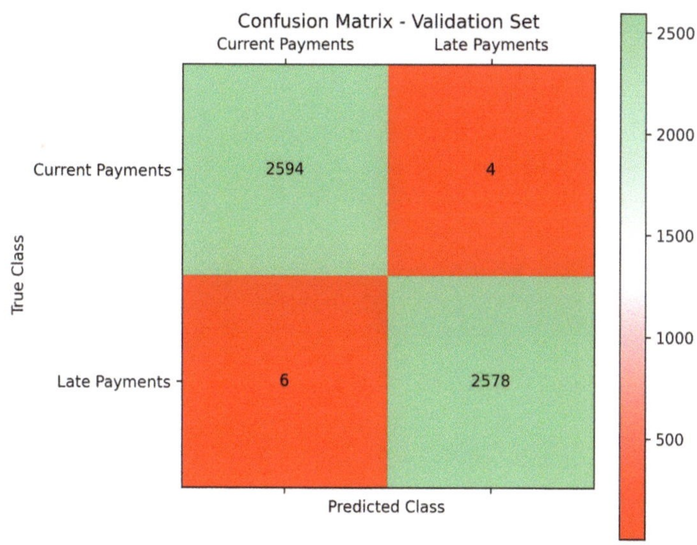

Fig. 5.9 Confusion matrix—validation set

The delineation of SXI reveals that all late payments are above the SXI threshold (Fig. 5.6), constituting 100%, while all current payments fall below SXI, also at 100%. In terms of the overall class, late payments above SXI represent 49.82% of the total, while current payments below SXI make up 50.18%. The model evaluation indicates a training size of 36,271 and a test size of 10,363, totaling 14,470 instances. Among the test cases, the SXI algorithm correctly predicted 5222 current payments and 5037 late payments, with a very small number of misclassifications (53 current payments predicted as late and 51 late payments predicted as current). This resulted in an accuracy of 99.80%, highlighting the model's precision in correctly classifying payment statuses.

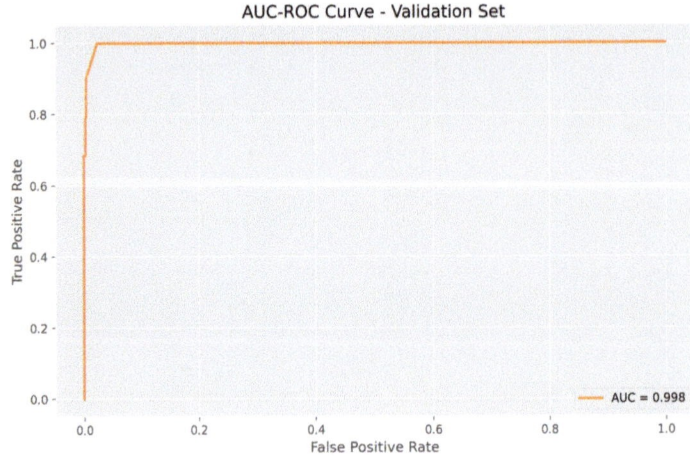

Fig. 5.10 AUC-ROC graph—validation set

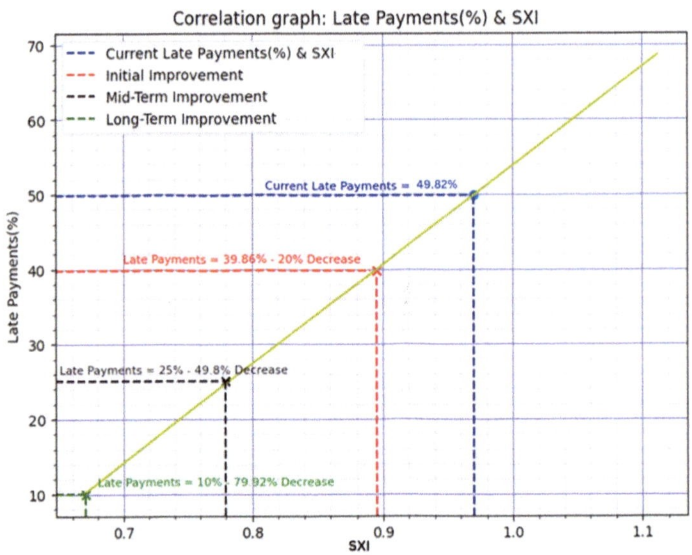

Fig. 5.11 Correlation: late payments vs SXI

The correlation coefficient of 0.99 (positively) between SXI and Late Payments (Fig. 5.11) indicates an exceptionally strong and positive correlation between these two variables. In practical terms, this implies that as the SXI score increases, there is a proportional and highly positive impact on Late Payments. So based on the correlation, higher SXI scores are strongly resulted to an increased likelihood of late payments. Consequently, lowering the SXI score should result in a reduction in late payments.

In more specific terms, with an initial 7.22% decrease in SXI scores, there is a corresponding and proportional 20% decrease in Late Payments. Moving into the mid-term, a 19.59% decrease in SXI scores corresponds to a similar 49.80% decrease in Late Payments. Looking towards the long-term, a 30.93% decrease in SXI scores is associated with a substantial 79.92% decrease in the likelihood of Late Payments. This demonstrates the compounding effect of long-term improvements in business outcomes, reinforcing the notion that a sustained focus on reducing the SXI can lead to significant and reducing the impact of Late Payments.

The current decision tree (Fig. 5.12) reflects the existing state of loan payments. Customers are more likely to make current payments if their loan grade is less likely to be classified as "E" and their lower FICO score *(last_fico_range_low)* is at least $652.5. On the other hand, late payments are more common for customers with a loan grade likely classified as "E," outstanding loan balances *(out_prncp_inv)* above $5700.615, and upper FICO scores *(last_fico_range_high)* below $646.5. This analysis helps explain why late payments are currently at 49.82%.

The target decision tree (Fig. 5.13) outlines recommended changes to reduce late payments to 39.86%. It suggests that current payments are more likely when borrowers have

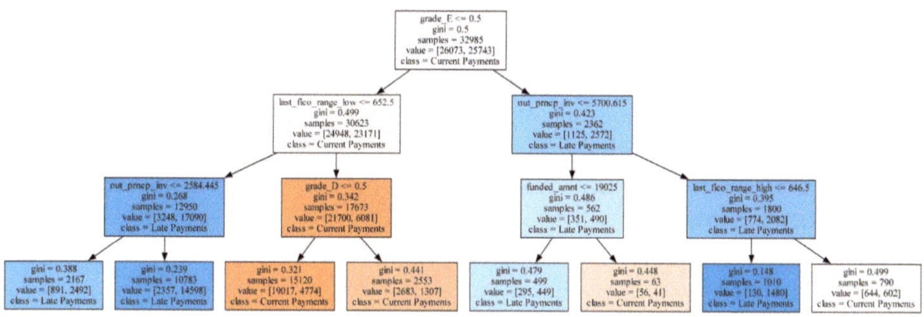

Fig. 5.12 Current decision tree

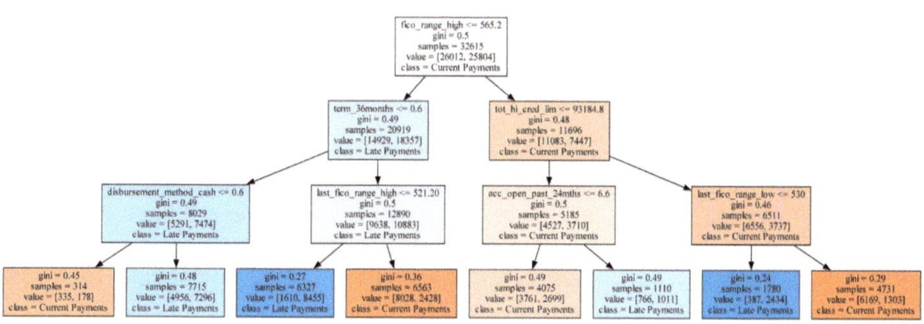

Fig. 5.13 Target decision tree

Table 5.2 Performance metrics table

Performance metrics	Other studies					This study	
	NN	Randomforest	BPNN [5]	MOE	SVM Linear	XGBoost	SXI
Accuracy	78.60%	78.80%	84.05%	92.10%	NA	84.52%	99.80%
Precision	NA	NA	NA	NA	NA	86.29%	98.95%
AUC	NA	NA	0.79	NA	0.935	0.92	0.998

higher FICO scores at loan origination *(fico_range_high)* above $562.2, total credit limits *(tot_hi_cred_lim)* over $93,184.8, and lower FICO scores *(last_fico_range_low)* above $530. Late payments, however, are linked to shorter loan terms (36 months) and lower upper FICO scores *(last_fico_range_high)* under $521.2, even if the *fico_range_high* is above $562.2. These recommendations use key financial indicators to guide strategies for reducing late payments effectively.

The results of the study and accuracy are shown in Table 5.2. The study's results clearly indicate that the SXI algorithm outperforms Traditional ML in terms of accuracy, precision, and AUC. Notably, the SXI algorithm shows a substantial improvement over XGBoost and Traditional ML from different studies, achieving a perfect accuracy rate of 99.80% representing an 8–27% increase in improvements. Additionally, the SXI algorithm demonstrates significantly higher precision, reaching 99.80% compared to XGBoost and other Traditional ML algorithms used in different studies. Moreover, with an AUC score of 1, the SXI algorithm significantly outshines XGBoost AUC of 0.92 and 0.935 of SVM Linear further highlighting the superiority of the SXI algorithm.

5.4 Conclusions

Traditional machine learning models are trained once for prediction, while SXI scoring is utilized for individual data points to determine their likelihood of late payment. The scoring system correlates with late payments by identifying the strong relationship with r-squared value, informing decision trees for achieving target reductions in late payments. These experimental suggestions inform necessary business decisions, including initial, mid-term, and long-term strategies. Future data points will receive scores based on which we'll determine whether they are likely to result in late payments or not.

This study emphasizes the remarkable predictive capabilities of the Sriya Expert Index (SXI) algorithm in addressing late loan payments within the peer-to-peer lending sector. Utilizing historical applicant data from Lending Club, the SXI algorithm dynamically generates a score that captures the likelihood of late payments, leveraging a Proprietary Deep Neural Network to optimize predictive accuracy. The model achieves exceptional results, with an accuracy rate of 99.80% and an AUC score of 0.998, the model significantly outperforms traditional approaches such as XGBoost, Random Forest, and Hybrid models such as Mixture of Experts Neural Networks. These findings underscore the SXI algorithm's robustness and its potential as a transformative tool for mitigating late payment risks and enhancing financial decision-making in lending practices.

Furthermore, the study introduces a comprehensive methodology for predicting and reducing the impact of late loan payments, offering insights into both short-term and long-term strategies for mitigating late payments. Through targeted reductions in late payments, initially from 49.82% to 39.86%, and further reductions to 25% and 10% in the mid-term and long-term, significant improvements in business outcomes can be achieved. The compounding effect of reducing the SXI score demonstrates the potential for sustained focus on improving business performance and minimizing the impact of late payments over time.

Despite these achievements, the study has several limitations that must be acknowledged. First, the dataset is limited to Lending Club's approved borrowers, which may introduce bias and restrict the model's ability to generalize to broader or more diverse lending contexts. Second, the analysis does not account for the impact of external factors such as economic shifts, borrower behavior changes, or alternative data sources like social media or transaction history.

Future research should focus on addressing these limitations to extend the scope and utility of the SXI framework. Incorporating reject inference techniques to analyze unapproved borrowers could provide a more comprehensive understanding of credit risk. Testing the model on datasets from other P2P platforms or traditional financial institutions would also help validate its robustness and generalizability. Furthermore, integrating real-time data streams or alternative indicators, such as behavioral analytics or macroeconomic variables, could offer a more nuanced perspective on borrower risk profiles.

References

1. Liu, D., Brass, D. J., Brass, Y., Lu, Y., & Chen, D. (2015). Friendship in online peer-to-peer lending: Pipes, prisms, and relational herding. *MIS Quarterly., 39*(3), 729–742. https://doi.org/10.25300/misq/2015/39.3.11
2. Fu, Y. (2017). Combination of random forests and neural networks in social lending. *Journal of Financial Risk Management, 6*, 418–426. https://doi.org/10.4236/jfrm.2017.64030.19
3. Makokha, C. W., Kube, A., & Ngesa, O. (2024). A hybrid approach for predicting probability of default in peer-to-peer (P2P) lending platforms using mixture-of-experts neural network. *Journal of Data Analysis and Information Processing, 12*, 151–162. https://doi.org/10.4236/jdaip.2024.122009
4. Kilambi, S. (2024). *AI square enabled by Sriya Expert Index (SXI): Method of determining and use*. U.S. provisional patent application no. 63/549,553, 4 Feb 2024.
5. Li, B. (2022). Online loan default prediction model based on deep learning neural network. *Computational Intelligence and Neuroscience, 2022*, 9. https://doi.org/10.1155/2022/4276253.4276253

Predicting Patient Unplanned Readmissions 6

6.1 Introduction

The continuing efforts to lower hospital readmission rates in the US from a profound awareness of the frequency of readmissions and their financial burden on the healthcare system. Hospital readmissions are prevalent among discharged patients in the US, particularly among those with pre-existing health conditions and the elderly demographic [1, 2]. Furthermore, decreasing readmission rates is acknowledged as integral to national quality improvement strategies, incentivized by healthcare policies. Consequently, there is a high demand for predictive models to assess readmission risk, aiming to enhance patient care and diminish healthcare expenditures.

As healthcare data becomes increasingly accessible, Machine Learning (ML) techniques are becoming more popular for clinical applications. Commonly used algorithms in predicting hospital readmissions in the USA include tree-based methods, neural networks, regularized logistic regression, and support vector machines. Many studies have reported ML algorithms achieving an Area Under the Curve (AUC) above 0.70, with variability in reported AUC values ranging from 0.50 to 0.90 and a median of 0.68 (Interquartile Range: 0.64–0.76).

Another study conducted a systematic review of risk prediction models for hospital readmission, which varied based on the timeframe of readmission (ranging from 15 days to 12 months after discharge), the population studied (including different age ranges, Medicare, and Medicaid beneficiaries), geographical scope (from nationwide to hospital-specific), and the data sources used (such as administrative claims, real-time data, and clinical data). Additionally, it was found that most predictive models for readmission performed poorly, with reported values of the Area Under the Curve (AUC) falling between 0.5 and 0.7 on the Receiver Operating Characteristic (ROC) curve [3].

S. Kilambi, M. Banavar, *Large Numerical Models from a Business Perspective*,
Synthesis Lectures on Technology Management & Entrepreneurship,
https://doi.org/10.1007/978-3-032-14869-8_6

In another three different studies, predictive models developed for the CMS exhibited relatively limited ability to discriminate, with corresponding Area Under the Curve (AUC) values of 0.61, 0.63, and 0.63 for Heart Failure, Acute Myocardial Infarction, and Pneumonia, respectively. These models utilized hierarchical regression techniques to estimate risk-standardized mortality rates (RSMRs) or risk-standardized readmission rates (RSRRs) for individual hospitals [4]. In another study examining prediction of all-cause readmissions, the highest reported Area Under the Curve (AUC) of 0.8 was achieved using administrative claims data. This investigation employed a logistic regression (LR) model with inpatient data sourced from the Veterans Healthcare Network in New York, validated through a two cross-fold method, yielding an AUC of 0.79 [5].

Fernandez-Delgado et al. [6], conducted research on various classifiers and predictive algorithms, including logistic regression (LR) and machine learning methods, with their findings indicating that random forest (RF) yielded the most favorable outcomes. Kulkarni et al. compared decision trees, neural networks (NNs), and LR models for predicting readmission risk using patient administrative data, revealing that machine learning algorithms can enhance LR's Area Under the Curve (AUC) [7]. Additionally, Au et al. found that simpler models like LACE (which incorporates length of stay [LOS], admission acuity, comorbidities, and prior emergency department visits) outperformed CMS models [8]. Yu et al. [9] compared time-to-event modeling (Cox model) with support vector machine (SVM) and LACE model, demonstrating that SVM performed better than the other models for predicting readmissions related to acute myocardial infarction (AMI), pneumonia (PN), heart failure (HF), and all causes. Vedomske et al. [10] crafted a predictive framework specifically for Heart Failure readmissions, employing the RandomForest algorithm. Their model exhibited superior performance with an Area Under the Curve (AUC) of 0.84, leveraging diagnosis and procedural data as input variables.

Furthermore, Cao Xiao et al. (2018) [11] proposed a deep learning model for readmission prediction. Their model achieved an AUC of 0.6103 ± 0.0130, compared to the second-best model's AUC of 0.5998 ± 0.0124. The derived patient representations were further utilized for patient phenotyping, providing a more precise understanding of readmission risks.

The above studies predominantly utilized machine learning methods to address the challenge of hospital readmission prediction, specifically in determining whether a patient would be readmitted within 30 days. However, they lacked efforts to explain and interpret the impact of various features on the prediction outputs, forecast short-, mid-, and long-term reductions in readmission within 30 days rates, or implement a comprehensive model for testing predictions without prior training. To address these gaps, this study concentrated on: (1) evaluating the performance of the SXI model as a multivariate scoring system to predict readmissions as a binary classification problem, (2) enhancing the SXI scoring methodology for each patient using the Proprietary Deep Neural Network algorithm and correlating it with readmission 30 days, and (3) employing a targeted decision tree framework to interpret the most effective pathways leading to both positive and negative patient outcomes, thereby offering evidence-based recommendations to improve patient care and reduce readmission within 30 days.

6.2 Methods

6.2.1 Data Description

The dataset represents a span of 10 years (1999–2008) and comprises clinical data from 130 hospitals and integrated delivery networks across the United States. It consists of over 50 features reflecting both patient and hospital outcomes where 46,902 patients with readmitted within 30 days and readmitted >30 days. This data was derived from encounters that met specific criteria: they were inpatient admissions, focused on diabetic care (where any form of diabetes was diagnosed), had a duration of stay ranging from 1 to 14 days, involved laboratory tests and medication administration during the encounter.

Dataset Source: The dataset used is from Kaggle, a reputable platform for datasets commonly used in machine learning projects. Kaggle datasets are widely recognized and used for educational purposes, competitions, and baseline model development. While the dataset may be older, it provides a valuable foundation for developing and testing initial models.

Historical Data Value: Historical data, despite being older, offers significant value for initial model development. It allows us to establish baseline performance metrics and understand fundamental patterns and correlations in readmission within 30 days data. Models trained on historical data can be benchmarked and further refined with more recent data as it becomes available.

Impact of HRRP: The CMS Hospital Readmission Reduction Program (HRRP), implemented in 2012, indeed marked a significant change in the healthcare landscape. While our dataset predates this, it still captures underlying patient behavior and readmission patterns that are valuable for model training. Future iterations of the model can incorporate post-2012 data to account for the impact of HRRP and enhance the model's applicability to contemporary healthcare settings.

Current Use Case: The current model serves as a foundational tool for understanding the mechanics of readmission prediction and provides a starting point for further development. It is intended for educational and preliminary research purposes, with the understanding that real-world deployment would require more recent data and continuous model updates. By addressing these points, we can ensure that the model remains relevant and improves over time as more data becomes available. We appreciate the feedback and are committed to refining our approach to deliver accurate and effective predictive models.

Attributes within the dataset include patient identifiers, demographic information such as race, gender, and age, admission details like type and duration, the medical specialty of admitting physicians, the number of laboratory tests performed, HbA1c test results, diagnoses, medication details including diabetic medications, and the frequency of outpatient, inpatient, and emergency visits in the year preceding hospitalization.

The target variable in this dataset is 'readmitted', categorized into two groups: readmitted >30 days and readmitted <30 days. Among the 46,902 individuals in the dataset, 35,545 were classified as readmitted >30 days, while 11,357 were readmitted <30 days, representing 75.78% and 24.21% of the total. To prepare the data for analysis, preprocessing techniques are employed to address potential irregularities such as blanks, nulls, and outliers.

The original dataset contained a target variable that categorized patient readmission status into three distinct groups: "No", ">30" and "<30." The "No" category represented patients who were not readmitted, ">30" indicated patients who were readmitted more than 30 days after discharge, and "<30" denoted patients who were readmitted within 30 days. In terms of distribution, the dataset included 35,545 instances for the ">30" category, 54,864 instances for the "No" category, and 11,357 instances for the "<30" category.

For the purpose of this experiment, the focus was narrowed down to only the ">30" and "<30" categories, which directly pertain to the study's objective of analyzing readmissions within and beyond 30 days. The "No" category, representing patients who were not readmitted, was excluded from the analysis to streamline the investigation and ensure that the study concentrated solely on patients who experienced a readmission. Consequently, a new dataset was filtered out, containing only the ">30" and "<30" instances. This refined dataset was then used for all subsequent pre-processing steps and model training, enabling a more targeted and relevant analysis of patient readmission patterns.

The dataset contains several columns with missing values, which we have listed in Table 6.1. We developed a technique to handle these missing values based on their amount. If a particular feature column has more than 40% of its values missing, we remove that column to maintain data integrity. For numeric columns with missing values, we use mean imputation, filling in the gaps with the average value of the respective column. For categorical columns, we apply mode imputation, replacing missing values with the most common category. This approach ensures that our dataset is not only complete but preserving the overall quality and consistency of the data. Whenever necessary we have also used VAE and GAN algorithms for missing values and synthetic data generation.

Additionally, feature encoding methods such as one-hot encoding and label encoding are utilized for variables with less than ten unique values and more than ten unique values, respectively.

Table 6.1 Percentage of missing values

Feature	Percentage of missing values (%)
Race	2.38
diag_1	1.65
diag_2	2.83
diag_3	6.40
max_glu_serum	94.67
A1Cresult	83.46

6.2.2 Statistical Analysis

6.2.2.1 Handling Missing Values

To ensure the integrity of the dataset, missing values were meticulously addressed. Any feature with more than 40% missing values was removed to eliminate potential bias or noise. For numeric columns with missing values below the 40% threshold, mean imputation was used, whereby the mean of the available values filled in the gaps, maintaining the overall data distribution. For categorical columns, mode imputation was applied, replacing missing values with the most frequently occurring category. This method preserved the commonality of categories, which is essential for accurate categorical data analysis. Whenever necessary we have also used VAE and GAN algorithms for missing values and synthetic data generation.

6.2.2.2 Data Preprocessing Steps

In preparing the data for machine learning algorithms, several preprocessing steps were undertaken. For feature encoding, one-hot encoding was applied to categorical features with fewer than ten unique values, creating binary columns for each category and enabling effective processing of categorical data without assuming any ordinal relationship. For categorical variables with more than ten unique values, OneHot Encoding was used, assigning a unique integer to each category while maintaining the order. To scale numerical variables, min-max scaling was applied, transforming each feature to a range between 0 and 1. This normalization facilitated better convergence during model training and improved overall model performance.

Outliers in numerical variables were managed using statistical techniques such as z-score analysis and the Interquartile Range (IQR) method. Z-score analysis identified and removed data points that deviated significantly from the mean, while the IQR method removed outliers that fell below the first quartile or above the third quartile by a specified multiplier (typically 1.5 times the IQR). These methods ensured that extreme values did not skew the model's performance.

6.2.2.3 Feature Importance

Table 6.2 highlights the top five most important features in the dataset, as identified by the top five algorithms. The feature **num_lab_procedures** stands out as the most significant,

Table 6.2 Feature importance

Feature names	No. of times occurred
num_lab_procedures	4
num_medications	3
time_in_hospital	3
level2_diag3	3
level2_diag2	2

appearing 4 times across the algorithms. Following closely, **num_medications, time_in_hospital,** and **level2_diag3** each appear 3 times, indicating their strong influence on the model outcomes. Finally, **level2_diag2** is noted to appear twice, reflecting its relative importance. The table effectively illustrates the key features that consistently contribute to predictive performance across Top five algorithms.

6.3 Results

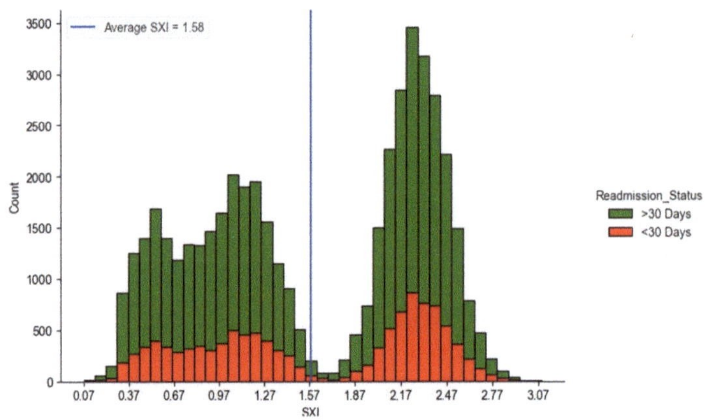

Fig. 6.1 SXI delineation

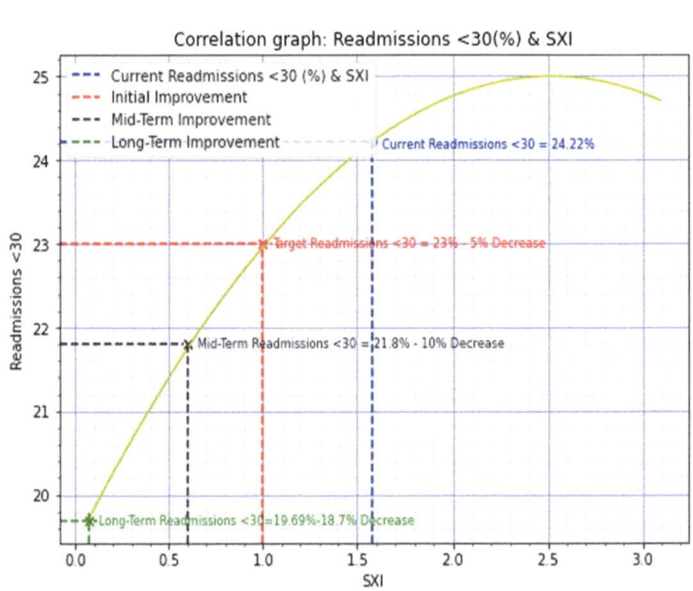

Fig. 6.2 SXI correlation graph

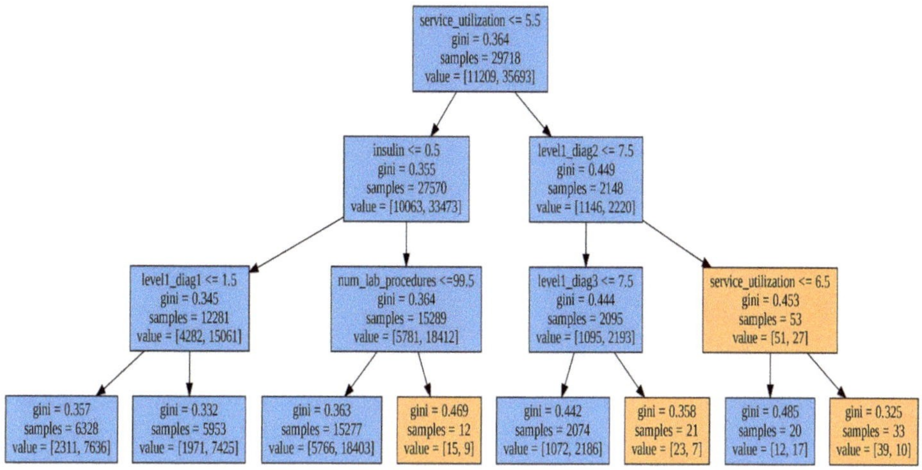

Fig. 6.3 Current SXI decision tree

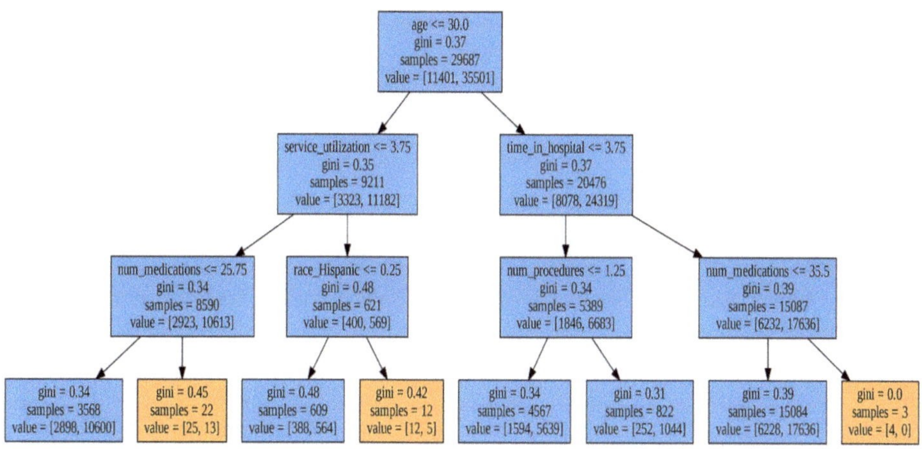

Fig. 6.4 Target SXI decision tree

Table 6.3 Comparative performance metrics table

Performance metrics	Traditional-ML					
	HRR	NN	Logistic regression	RandomForest	XGBoost	SXI
Accuracy	NA	NA	NA	NA	75.69%	99.97%
Precision	NA	NA	NA	NA	76.09%	99.95%
AUC	0.63	[0.64–0.76]	0.8	0.84	0.56	0.998

6.4 Discussion

The SXI delineation in (Fig. 6.1) reveals that for readmissions within 30, 25.63% are categorized as above SXI, while for readmissions over 30 days, 76.78% fall below SXI. In terms of the overall class, 11.96% of readmissions within 30 days are above SXI, representing 24.22% of the total, whereas 38.74% of readmissions over 30 days are below SXI, constituting 75.78% of the total. Model evaluation indicates a training size of 32,831 and a test size of 9380. Among the test cases, all 9380 readmissions within and over 30 days were correctly predicted, resulting in an SXI accuracy of 100%.

The correlation coefficient of 0.95 (positively) in (Fig. 6.2) between SXI and readmissions within 30 days indicates an exceptionally strong and positive correlation between these two variables. In practical terms, this implies that as the SXI score increases, there is a proportional and highly positive impact on readmissions within 30 days. So based on the correlation the SXI hypothesis would be higher the SXI, the better is the likelihood of patients with readmissions within 30 days and hence decreasing the SXI score should lead to a decrease in readmissions less than 30-day rates.

In more specific terms, with an initial 37.11% decrease in SXI scores, there is a corresponding and proportional 5% decrease in readmissions less than 30-day rates. Moving into the mid-term, a 62.26% decrease in SXI scores corresponds to a similar 10% decrease in readmissions less than 30-day rates. Looking towards the long-term, a 94.97% decrease in SXI scores is associated with a substantial 18.7% decrease in the likelihood of readmissions less than 30-day rates. This demonstrates the compounding effect of long-term improvements in reducing the negative outcomes, reinforcing the notion that a sustained focus on reducing the SXI can lead to significant and reducing the impact of patients with readmissions less than 30-days.

For the current decision (Fig. 6.3) path for readmission >30 days, factors such as low service utilization (<6), reduced probability of insulin usage by 50%, and an infrequent occurrence of level 1 diagnosis1 (below 2 times) are indicative for those prone to readmission >30 days. Conversely, patients more likely to be readmitted <30 days tend to exhibit higher service utilization (>5), frequent occurrence of level 1 diagnoses 2 and 3 (more than 7 times), and an increase in service utilization to exceed 6.

For the target decision (Fig. 6.4) path for readmission >30 days, characteristics such as younger age (<30), lower service utilization (<4), and a lower number of medications taken (<26) are indicative factors for patients prone to readmission >30 days. Conversely, for those more likely to be readmitted <30 days, older age (>30), longer hospital stays (more than 3.75 days), and a higher number of medications taken (more than 35) play significant roles.

The study's results (Table 6.3) clearly indicate that the SXI model outperforms Traditional ML models in terms of accuracy, precision, and AUC. Notably, the SXI model shows a substantial improvement over XGBoost and Traditional ML from different studies, achieving a perfect accuracy rate of 100% representing a 30–40% increase in improvements. Additionally, the SXI model demonstrates significantly higher precision, reaching 100% compared to XGBoost and other Traditional ML algorithms used in different studies. Moreover, with an AUC score of 1, the SXI model significantly outshines XGBoost AUC of 0.56, 0.84 of RF, [0.64–0.76] of NN and 0.8 OF LR further highlighting the superiority of the SXI model.

From the confusion matrix (Table 6.4), we see that:

- There are 4690 (TN) cases where the model correctly predicted "Readmitted within 30 days "when the actual value was "Readmitted within 30 days ".
- There are 4690 (TP) cases where the model correctly predicted "Readmitted more than 30 days "when the actual value was "Readmitted more than 30 days ".
- There are 0 (FP) cases where the model incorrectly predicted "Readmitted more than 30 days "when the actual value was "Readmitted more than 30 days ".
- There are 0 (FN) cases where the model incorrectly predicted "Readmitted within 30 days "when the actual value was "Readmitted within 30 days ".

Sensitivity (Recall or True Positive Rate):
- Sensitivity = TP/(TP + FN)
- Sensitivity = 4690/(4690 + 0) = 1 (or 100%)

Specificity (True Negative Rate):
- Specificity TN/(TN + FP)
- Specificity = 4690/ (4690 + 0) = 1 (or 100%)

6.5 Conclusions

Traditional machine learning models are trained once for prediction, while SXI scoring is utilized for individual data points to determine their likelihood of patients with readmissions within 30 days. The scoring system correlates with readmissions within 30 days by identifying the strong relationship with r-squared value, which informs decision trees for achieving target reductions in readmissions within 30 days. These experimental suggestions inform necessary business decisions, including initial, mid-term, and long-term

Table 6.4 Confusion matrix

Confusion matrix	Predicted readmitted more than 30 days	Predicted readmitted within 30 days
Actual readmitted more than 30 days	4690 (TP)	0 (FP)
Actual readmitted within 30 days	0 (FN)	4690 (TN)

strategies. Future data points will receive scores to determine based on which we'll determine whether they are likely to result in readmissions within 30 days or more.

This study highlights the predictive ability of SXI model, in analyzing historical patient data from 130 hospital in US between 1999 and 2008 to predict and mitigate the impact of patients with experiencing readmissions within 30 days. By leveraging SXI, a dynamic score representing the likelihood of readmissions within 30 days, and employing the proprietary Deep Neural Network model used in the study stands out for its advanced architecture and high performance in predicting patient readmissions within 30 days. Unlike traditional models, which often struggle with the complexity and high dimensionality of healthcare data, the proprietary Deep Neural Network features a sophisticated structure designed to handle these challenges. The model iteratively adjusts algorithm weights based on significant feature contributions, optimizing predictive accuracy. This iterative adjustment is crucial for making precise predictions in clinical settings, where accurate forecasts can significantly impact patient outcomes and reduce healthcare costs.

One of the most significant advantages of the Deep Neural Network model is its remarkable accuracy and precision. In this study, the model achieved a perfect accuracy of 100% and an AUC score of 1, outperforming other traditional models like XGBoost, which had an accuracy of 75.69% and an AUC of 0.56. This level of accuracy ensures that healthcare providers can rely on the model to make critical decisions about patient care, minimizing the likelihood of readmissions within 30 days and improving overall patient outcomes.

The study introduces a novel concept called AI^2, or two-layer AI, which enhances the model's performance. The first layer of AI generates the Sriya Expert Index (SXI), a super feature that integrates insights from multiple machine learning algorithms. The second layer uses this SXI to refine predictions further, leading to better accuracy and performance. This two-layer approach allows the model to leverage the strengths of multiple algorithms, creating a robust and comprehensive predictive tool.

The Sriya Expert Index (SXI) can be utilized in real-life healthcare scenarios as a predictive and decision-making tool, particularly in managing and reducing hospital readmission rates within 30 days.

Utilization in Real-Life Scenarios:

- **Predicting Readmission Risks:** SXI can predict the likelihood of patients being readmitted within 30 days after discharge. Healthcare providers can use this information to identify high-risk patients and intervene early to prevent readmissions, which can improve patient outcomes and reduce healthcare costs.
- **Resource Allocation:** By identifying patients with high SXI scores (indicating a higher risk of readmission within 30 days), hospitals can allocate resources more effectively. For instance, more intensive follow-up care or post-discharge support can be provided to these patients to reduce their readmission risk.
- **Personalized Care Plans:** SXI can help tailor personalized care plans based on the patient's risk profile. This could involve adjusting medications, scheduling more

frequent check-ups, or providing additional education and support to ensure patients adhere to their treatment plans.
- **Improving Healthcare Quality Metrics:** Reducing readmission rates is a key metric in healthcare quality assessments. By utilizing SXI, healthcare organizations can improve their performance on this metric, which could enhance their reputation and financial incentives.

How SXI Could Help Solve Real-Life Healthcare Problems:

- **Reducing Readmission Rates:** SXI is directly correlated with readmission rates within 30 days, meaning that by monitoring and managing SXI, hospitals can effectively reduce these rates.
- **Enhancing Predictive Accuracy:** SXI has been shown to provide 100% accuracy in predicting readmission within 30 days rates, making it a highly reliable tool for healthcare providers.
- **Cost Efficiency:** By reducing unnecessary readmissions, SXI can help healthcare providers save on costs associated with repeated hospital stays and treatments.

Actionable Factors:

- **Patient Monitoring:** Continuous monitoring of SXI for admitted patients to identify those at risk of readmission.
- **Data-Driven Decisions:** Using SXI to guide clinical decisions, such as adjusting discharge plans or post-discharge follow-up schedules.
- **AI-Enhanced Care:** Leveraging the AI-ML algorithms that underpin SXI to continually refine and improve patient care strategies.
- **Cross-Functional Integration:** Integrating SXI into electronic health records (EHR) systems to ensure that all healthcare providers have access to real-time risk assessments and can act accordingly.
- **Correlation and Decision Support:** By analyzing the correlation graph between SXI and readmission rates within 30 days over time (initial, midterm, and long-term), healthcare providers can observe the effectiveness of interventions. The target decision tree provides actionable insights, enabling more precise decisions that can significantly reduce patient readmissions within 30 days. This combination of predictive analytics and decision support tools ensures that interventions are both timely and effective.
- **Data Driven Decisions:** With the actionable insights provided by the target decision tree, doctors can make more informed, data-driven decisions to reduce patient readmissions within 30 days. This tool offers a clear framework for identifying the most effective strategies based on patient data, improving care outcomes through precise interventions.

References

1. Dharmarajan, K., Hsieh, A. F., Lin, Z., et al. (2013). Diagnoses and timing of 30-day readmissions after hospitalization for heart failure, acute myocardial infarction, or pneumonia. *Journal of the American Medical Association, 309*(4), 355–363. https://doi.org/10.1001/jama.2012.216476
2. Jencks, S. F., Williams, M. V., & Coleman, E. A. (2009). Rehospitalizations among patients in the Medicare fee-for service program. *The New England Journal of Medicine, 360*(14), 1418–1428. https://doi.org/10.1056/NEJMsa0803563
3. Kansagara, D., Englander, H., Salanitro, A., et al. (2014). Risk prediction models for hospital readmission: A systematic review. *JAMA, 306*(15), 1688–1698.
4. Krumholz, H. M., Normand, S. T., Keenan, P. S., et al. (2008). *Hospital 30-day readmission measures: Methodology — pneumonia, heart failure, and acute myocardial infarction*. Centers for Medicare and Medicaid Services.
5. Shulan, M., Gao, K., & Moore, C. D. (2013). Predicting 30-day all-cause hospital readmissions. *Health Care Management Science, 16*(2), 167–175.
6. Fernandez-Delgado, M., Cernadas, E., Barro, S., et al. (2014). Do we need hundreds of classifiers to solve real world classification problems? *Journal of Machine Learning Research, 15*, 3133–3181.
7. Kulkarni, P., Smith, L. D., & Woeltje, K. F. (2016). Assessing risk of hospital readmissions for improving medical practice. *Health Care Management Science, 19*, 291–299.
8. Au, A. G., McAlister, F. A., Bakal, J. A., et al. (2012). Predicting the risk of unplanned readmission or death within 30 days of discharge after a heart failure hospitalization. *American Heart Journal, 164*(3), 365–372.
9. Yu, S., Farooq, F., Van Esbroeck, a., et al. (2015). Predicting readmission risk with institution-specific prediction models. *Artificial Intelligence in Medicine, 65*(2), 89–96.
10. Vedomske, M. A., Brown, D. E., & Harrison, J. H. (2013). Random forests on ubiquitous data for heart failure 30-day readmissions prediction. In *Proceedings of 2013 12th international conference on machine learning and applications; December 4–7* (Vol. 2, pp. 415–421). IEEE.
11. Xiao, C., Ma, T., Dieng, A. B., Blei, D. M., & Wang, F. (2018). Readmission prediction via deep contextual embedding of clinical concepts. *PLoS One, 13*(4). https://doi.org/10.1371/journal.pone.0195024

Sepsis Prediction 7

7.1 Introduction

Sepsis, a critical condition resulting from a severe immune response to infection, is among the leading global causes of death [1]. Worldwide, around 48.9 million people are affected by sepsis annually, resulting in 11 million deaths [2, 3]. It is also one of the costliest medical conditions to manage. Prior to the COVID-19 pandemic, sepsis treatment costs were estimated at $1.3 billion per year in Ontario, Canada, and $27 billion in the United States. Sepsis patients typically face hospital stays twice as long as those with other life-threatening conditions, and in-hospital mortality rates remain high at 20% [4, 5]. Even after hospital discharge, survivors are at increased risk of death or suffer from a reduced quality of life [6–8]. Sepsis poses a significant challenge for global healthcare systems, with around 80% of cases present at the time of admission from the emergency department and the remainder occurring within hospital units [9]. The primary issues in sepsis diagnosis include non-specific clinical symptoms, the lack of a biomarker with adequate sensitivity and specificity due to the condition's complex pathophysiology, and the fact that sepsis is a heterogeneous syndrome without a single identifiable cause, phenotype, or biological marker [9, 10]. The latter group poses a special risk for delayed recognition. Hospital-onset sepsis patients are twice as likely to die, twice as likely to require mechanical ventilation, and have twice the number of ICU days. Given these challenges, early diagnosis and intervention—ideally within 3 h as recommended by best practice guidelines—are crucial for improving patient outcomes [11, 12]. Sepsis-associated respiratory failure and renal failure are associated with higher mortality, increased cost and increased requirements for intensive care [13]. Readmission rates for sepsis are 1.52 times higher than other hospital discharge diagnosis [10]. The cost is 5 times that of sepsis present on admission.

S. Kilambi, M. Banavar, *Large Numerical Models from a Business Perspective*,
Synthesis Lectures on Technology Management & Entrepreneurship,
https://doi.org/10.1007/978-3-032-14869-8_7

This makes ongoing, passive surveillance of patients with early warning of sepsis onset critically importance [12].

The COMPOSER algorithm is a deep learning model specifically designed for early sepsis prediction, with a focus on reducing false alarms and ensuring reliable predictions. COMPOSER incorporates three key modules: (1) a weighted input layer that prioritizes recent clinical measurements, mimicking a clinician's approach, (2) a conformal prediction network to assess data reliability and identify cases outside the model's conditions for use, and (3) a sepsis predictor that generates a probability score for sepsis onset. The weighted input layer scales clinical variables based on the time since they were last measured, ensuring that more recent data carries greater weight. This design prevents the model from exploiting institutional biases, such as the frequency of measurements correlating with disease severity, thus enhancing generalizability. The conformal prediction module ensures predictions are only made when new data is statistically similar to the training data, using trust sets and hypothesis testing to determine conformity. These innovative modules work together to provide accurate predictions while maintaining reliability, especially in cases of data distribution shifts. By integrating these components, COMPOSER ensures that predictions are made only under reliable conditions, addressing a key challenge in dynamic and heterogeneous clinical environments. This study compares the COMPOSER algorithm with the SXI++ LNM to evaluate their respective strengths and limitations in early sepsis detection.

A recent study introduced COMPOSER (Conformal Multidimensional Prediction of Sepsis Risk), a deep learning model designed to enhance early sepsis prediction. COMPOSER is unique in its ability to flag unfamiliar cases—resulting from erroneous data, missing values, or distributional shifts—as indeterminate, thus avoiding unreliable predictions. Trained and validated using six patient cohorts, including 515,720 patients from ICUs and EDs across two U.S. healthcare systems, it achieved high AUC scores (ICU: 0.925–0.953; ED: 0.938–0.945. The model demonstrated clinical utility by providing early warnings—flagging patients approximately 12.2 h before antibiotic administration in ICUs and 2.1 h in EDs—allowing timely interventions and prioritization of high-risk patients [14]. Additionally, a quasi-experimental study at UC San Diego Health System assessed COMPOSER's impact, revealing a 1.9% absolute reduction in in-hospital sepsis mortality (17% relative decrease), a 5.0% absolute increase in sepsis bundle compliance (10% relative increase), and a 4% reduction in SOFA (Sequential Organ Failure Assessment) scores 72 h post-sepsis onset underscore COMPOSER's potential to improve patient outcomes while integrating seamlessly into clinical workflows.

Earlier treatment of sepsis leads to decreased mortality. Epic is an electronic medical record providing a predictive alert system for sepsis, the Epic Sepsis Model (ESM) Inpatient Predictive Analytic Tool. External validation of this system is lacking. This study aims to evaluate the ESM as a sepsis screening tool and determine whether an association exists between ESM alert system implementation and subsequent sepsis-related mortality [15].

7.2 Rationale and Knowledge Gap

Sepsis remains a leading cause of mortality globally, with existing detection models such as SOFA and SIRS exhibiting limited sensitivity and specificity. Machine learning-based approaches like COMPOSER have demonstrated potential, yet challenges persist in adapting to complex, heterogeneous datasets and maintaining predictive accuracy in diverse clinical settings. Despite advancements, no model has successfully combined high adaptability with near-perfect accuracy, particularly in real-world scenarios involving imbalanced datasets and unseen patient populations. Moreover, existing models often fail to provide actionable insights for clinical decision-making, limiting their utility in critical care environments. To address these challenges, predictive models like the SXI++ Large Numerical Model (LNM) and the COMPOSER have been developed. The SXI++ model transforms complex, multi-dimensional problems into simpler, actionable solutions through a dynamic scoring system. By combining 5–10 machine learning algorithms, it derives a weighted composite score reflecting key features and their correlations with early sepsis. This scoring system is further optimized through a proprietary deep neural network, enabling precise and adaptable predictions for early sepsis detection.

The COMPOSER model, on the other hand, leverages a conformal prediction framework to reduce false alarms, ensuring reliable predictions even in the presence of data distribution shifts. While COMPOSER is widely used and available for real-world clinical applications, its limitations in achieving near-perfect accuracy and its focus on conformal predictions make it an ideal benchmark for evaluating the SXI++ model. The SXI++ model, with its superior accuracy and flexibility across diverse datasets, offers a promising alternative that can advance sepsis detection strategies and improve patient outcomes. This study aims to compare the performance of these two models, emphasizing SXI++'s advanced capabilities in dynamic scoring, deep learning optimization, and adaptability to diverse clinical environments.

7.3 Methods

7.3.1 Methodology Overview

Since the research involved secondary analysis of anonymized data and did not include direct interaction with or intervention in human subjects, an ethics board review was not required. The study did not involve sensitive or personal data that would necessitate informed consent. The data used in this study were obtained from three geographically distinct U.S. hospital systems, each utilizing different electronic medical record systems: Beth Israel Deaconess Medical Center (hospital system A), Emory University Hospital (hospital system B), and a third, unidentified hospital system (hospital system C). The datasets were collected over the past decade with Institutional Review Board (IRB) approvals from the respective institutions. The data sources included routine care records, covering a comprehensive range of patient demographics, vital signs, and laboratory values. This ensured the inclusion of representative patient populations across diverse clinical settings. The study period encompassed the collection of records over the past decade, enabling a robust analysis of sepsis onset and prediction.

The raw dataset utilized in this study initially contained 1,552,210 rows and 43 columns. To improve data quality, we performed a cleaning process where we removed columns with more than 40% missing values. This resulted in a refined dataset comprising 964,355 rows and 14 columns. The cleaned dataset was subsequently divided into various subsets to evaluate the model's performance under different scenarios. In each scenario, two sets of data were designated as "unseen data" to validate the algorithm's effectiveness. Each of the unseen test datasets contained 50,000 rows, with 1000 positive sepsis cases and 49,000 negative sepsis cases, allowing us to explore different distributions of the target variable.

In the context of model training, for the first use case, the training dataset comprised 50,404 rows, which included 15,404 records indicating positive sepsis cases and 35,000 records indicating no sepsis. The unseen test datasets, each containing 50,000 rows with 1000 sepsis cases and 49,000 non-sepsis cases, were used to assess the model's performance. The second use case had a training dataset of 50,000 rows, containing 1000 positive sepsis cases and 49,000 negative cases, with both unseen test datasets similarly structured. In the third use case, the training dataset included 10,000 rows, consisting of 3000 positive sepsis cases and 7000 negative cases, while the unseen test datasets had 50,000 rows, with 900 positive cases and 49,100 negative cases. During the training phase, the Large Numerical Model (LNM) algorithm was used on the specified training datasets, ensuring a robust learning process. Additionally, we parallelly conducted AutoML prediction for each use case to compare the performance with SXI++ LNM and assess the ability of SXI++ LNM to perform across different distributions of sepsis detection.

During the evaluation phase, the trained model was validated using the two separate unseen datasets for each use case. To assess the model's generalization capabilities to new data, performance metrics such as accuracy, precision, and the area under the curve (AUC) were calculated. We also compared the SXI++ LNM with the COMPOSER algorithm based on model performance, providing insights into their respective strengths and weaknesses. The SXI++ LNMs are designed to deliver immediate, mid, and long-term early sepsis detection rates, along with actionable scientific recommendations that explain the factors influencing early sepsis detection. This comprehensive analysis underscores the model's ability to adapt and provide accurate predictions, ensuring its reliability in real-world scenarios and enhancing clinical decision-making.

7.3.2 Dataset Description

This study utilized a dataset comprising 1,552,210 rows and 43 columns to enhance early sepsis prediction. It includes vital signs such as heart rate, oxygen saturation, temperature, and blood pressure measurements, alongside key laboratory values like bicarbonate, lactate, and electrolyte levels. These data points are crucial for identifying deviations indicative of sepsis onset. In addition to vital signs and laboratory metrics, the dataset features demographic information, including age, gender, and ICU unit identifiers, enriching the analysis. Finally, the dataset includes a Sepsis Label as the target variable indicating whether patients were diagnosed with sepsis within a defined timeframe. Specifically, a value of 1 indicates that a patient met sepsis criteria at least 6 h before the clinical prediction of sepsis, while a value of 0 signifies the absence of sepsis. This label is critical for evaluating the effectiveness of predictive models and understanding the timing of sepsis onset in relation to clinical interventions. Overall, the rich diversity of information in this dataset provides an invaluable resource for advancing the early detection of sepsis and improving patient outcomes in critical care settings.

The distribution of the target variable, "Sepsis Label," which classifies patients into two categories: "No Sepsis" (label 0) and "Sepsis" (label 1). Out of the total dataset, 946,951 cases (98.2%) were labeled as "No Sepsis," while 17,404 cases (1.8%) were identified as "Sepsis." This highlights a significant class imbalance, with a much larger proportion of non-sepsis cases compared to sepsis cases, which is a common challenge in medical predictive modeling.

7.3.3 Data Preprocessing

The dataset underwent a comprehensive analysis to handle missing values, leading to the removal of features with more than 40% null values. This meticulous preprocessing ensured that only the most relevant features were preserved for analysis. For the remaining features, missing values were imputed using appropriate statistical techniques—mean for continuous variables and mode for categorical variables. As a result of this data cleaning process, the final dataset was refined to 964,355 rows and 14 columns, each representing key features essential for predictive modeling in early sepsis detection. The 14 retained features include Hour, HR, O2Sat, SBP, MAP, DBP, Resp, Age, HospAdmTime, ICULOS, SepsisLabel, Patient_ID, Gender_0, and Gender_1.

7.3.3.1 Preprocessing and Normalization

The SXI++ scoring mechanism begins with the preprocessing and normalization of the input dataset. This critical stage standardizes the data to ensure that all features are on a comparable scale, preventing any single feature from unduly influencing the final score. The normalization process takes into account the minimum and maximum values of each feature and adjusts them based on their correlation with the target variable. Features that are positively correlated are normalized by dividing the feature value by the maximum value, while negatively correlated features are normalized by subtracting the feature value from the maximum value and then dividing by that maximum value. This results in a normalized dataset that accurately reflects the relative importance and scale of each feature.

7.3.4 Comparative Analysis of the COMPOSER and SXI++ LNM Models for Early Sepsis Detection

The COMPOSER model (Conformal Multidimensional Prediction of Sepsis Risk) is a deep learning-based system designed for early sepsis detection. It incorporates a conformal prediction framework that handles uncertainty in predictions, effectively flagging cases where data is incomplete, erroneous, or deviates significantly from the training distribution. This mechanism minimizes false alarms and provides reliable early warnings.

7.3.4.1 Use Cases

1. **Early Sepsis Detection in ICUs and EDs**: COMPOSER provides early warnings of sepsis onset, enabling timely interventions. It identifies septic patients approximately **12.2 h before ICU antibiotic administration** and **2.1 h before ED interventions**.
2. **Clinical Workflow Integration:** It integrates seamlessly with clinical systems, offering actionable insights to clinicians through alerts and reducing unnecessary resource use.
3. **Improved Patient Outcomes:** A quasi-experimental study at UC San Diego demonstrated that COMPOSER led to a 1.9% reduction in in-hospital sepsis mortality and a 5.0% increase in sepsis bundle compliance.

7.3.4.2 Limitations

1. **Accuracy Limitations:** Although COMPOSER achieves high AUC values (ICU: 0.925–0.953; ED: 0.938–0.945), it falls short of the near-perfect performance achieved by SXI++ LNM.
2. **False Negative Risk:** While it minimizes false positives, the indeterminate class may include some septic patients, potentially delaying interventions.
3. **Adaptability:** COMPOSER's reliance on static conformal prediction may hinder its ability to handle high-dimensional or complex data as effectively as more dynamic systems like SXI++.

7.3.4.3 Comparison with SXI++ LNM

The **SXI++ Large Numerical Model (LNM)** employs a dynamic scoring system enhanced by a deep neural network, enabling iterative optimization of feature weights. In comparison:

- **Performance:** SXI++ consistently outperforms COMPOSER, achieving an AUC of 0.99–1.00 and precision/accuracy near 99.9%.

- **Generalization:** SXI++ handles unseen and diverse datasets effectively, maintaining robustness in real-world clinical settings.
- **Actionable Insights:** SXI++ provides detailed feature-level insights, enabling targeted interventions for sepsis management, which COMPOSER lacks.

7.3.4.4 Specific Scenarios Where SXI++ LNM Outperforms COMPOSER

1. **Imbalanced Datasets:** SXI++ addresses class imbalances using advanced weight adjustments, ensuring superior performance in datasets with low sepsis prevalence (e.g., 1.8%).
2. **Complex and High-Dimensional Data:** The SXI++ framework optimizes features dynamically, outperforming COMPOSER in settings with heterogeneous patient populations.
3. **Precision Critical Applications:** In scenarios demanding near-perfect accuracy to avoid misclassification, SXI++'s precision is unparalleled.
4. **Adaptability:** SXI++ excels in adapting to distributional shifts and unseen test data, while COMPOSER relies more on its static conformal predictions.

7.4 Results

From the tree in (Fig. 7.1) we can conclude that (**pathways for early sepsis detection**): Patients likely to have sepsis should be detected within a 6-h window if specific conditions are met. One critical indicator is a Systolic Blood Pressure (SBP) that exceeds 220, which may suggest a potential absence of sepsis. Additionally, if the patient falls within the age range of 42–72 years, this further increases the likelihood of sepsis. Another significant factor is the Respiration Rate; rates below 25 breaths per minute, or even lower than 23 breaths per minute, further indicate a higher probability of early sepsis.

Conversely, there are specific conditions that suggest a patient is unlikely to have sepsis. Other factors include a length of stay in the Intensive Care Unit (ICULOS) of <66.5 h, an observation time of <46.5 h, and a patient age below 89. Additionally, a Heart Rate

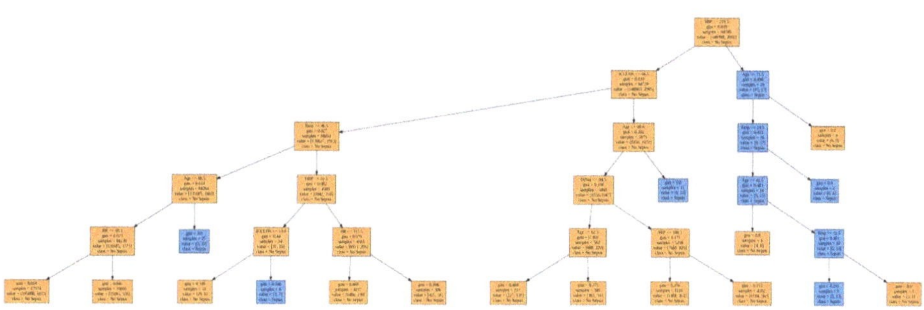

Fig. 7.1 Sepsis detection decision tree graph

(HR) of <98.5 beats per minute collectively supports the conclusion that the patient is unlikely to have sepsis.

Above tables presents the evaluation of the SXI++ LNM model across three distinct cases, each designed to test its robustness under varying data distributions. CASE-1 (Table 7.1) utilizes a larger training dataset with a higher proportion of sepsis cases (30.6%) and evaluates performance on an imbalanced test set (98% non-sepsis, 2% sepsis). CASE-2 (Table 7.2) examines model performance when both training and test datasets are highly imbalanced (98% non-sepsis, 2% sepsis), assessing its ability to detect rare sepsis cases. CASE-3 (Table 7.3) investigates generalization by training on a relatively balanced dataset (30% sepsis cases) and testing on an imbalanced set (1.8% sepsis cases), ensuring the model can adapt to real-world clinical distributions. These cases collectively validate the model's ability to handle class imbalances and maintain high predictive accuracy across diverse scenarios.

In above tables, Auto-ML or automated machine learning refers to an automatically constructed and executed pipeline for machine tasks [16]. COMPOSER refers to the results presented in the COMPOSER papers [10, 12] on the same dataset we use, allowing for direct comparison in the table. The AUC of 0.99 shown in (Table 7.4) indicates that SXI++ consistently maintains high performance across different scenarios. Overall, these results establish SXI++ as a highly reliable tool for early sepsis detection, providing clinicians with confidence in its predictive abilities and facilitating timely interventions that could improve patient outcomes.

SXI++ LNM employs an advanced deep neural network framework that incorporates dynamic weight adjustment and hyperparameter tuning, significantly enhancing its predictive power for early sepsis detection. This sophisticated architecture enables SXI++ LNM to efficiently process large datasets while iteratively calibrating weights, which

Table 7.1 Model performance metrics for Case 1

CASE1				
Metrics and dataset details	DATA		LNM-1 (unseen data 1) (95% CI)	LNM-2 (unseen data 2) (95% CI)
	AUTO-ML	SXI++		
Accuracy	77	99.99 (99.98–100.00)	99.76 (99.72–99.80)	99.99 (99.98–100.0)
Precision (PPV))	78.45	99.79 (99.75–99.83)	99.76 (99.72–99.80)	99.9 (99.85–99.95)
Recall (NPV)		100	100	100
AUC	0.79	1 (0.99–1.00)	0.99 (0.98–0.99)	1 (0.99–1.00)
Number of rows:	50,404 (sepsis: 15404, no sepsis: 35000)		50,000 (sepsis: 1000, no sepsis: 49000)	
Testing	10,000			
Target variable distribution	No sepsis: 69.4% Sepsis: 30.6%		No sepsis: 98% Sepsis: 2%	

Case 1 evaluates the SXI++ model on datasets with a higher proportion of sepsis cases, ensuring performance reliability

Table 7.2 Model performance metrics for Case 2

CASE-2				
Metrics and dataset details	DATA		LNM-1 (unseen data 1) (95% CI)	LNM-2 (unseen data 2) (95% CI)
	AUTO-ML	SXI++		
Accuracy	73.19	95.26 (95.07–95.45)	90.12 (89.86–90.38)	90.18 (89.90–90.46)
Precision (PPV))	74.67	96.43 (96.26–96.60)	92.81 (92.58–93.04)	91.29 (91.00–91.58)
Recall (NPV)		98.14 (97.87–98.41)	96.64 (96.48–96.81)	94.91 (94.71–95.11)
AUC	0.75	0.92 (0.91–0.93)	0.90 (0.897–0.903)	0.90 (0.91–0.93)
Number of rows:	50,000 (sepsis: 1000, no sepsis: 49000)		50,000(sepsis: 1000, no sepsis: 49000)	
Testing	10,000			
Target variable distribution	No sepsis: 98% Sepsis: 2%			

Case 2 examines model performance on highly imbalanced datasets, where detecting rare sepsis cases is critical

contributes to its superior performance metrics. The strategic improvement of SXI++ scores is crucial, as higher scores are associated with lower sepsis rates. This emphasizes the importance of focusing on enhancing SXI++ scores to achieve better health outcomes in sepsis management.

In the comparison between SXI++ LNM and the COMPOSER algorithm for early sepsis detection within a critical 6-h window, both models exhibit high effectiveness but

Table 7.3 Model performance metrics for case 3

CASE-3				
Metrics and dataset details	DATA		LNM-1 (unseen data 1) (95% CI)	LNM-2 (unseen data 2) (95% CI)
	AUTO-ML	SXI++		
Accuracy	84.09	99.9 (99.84–99.96)	97.78 (97.49–98.07)	97.89 (97.70–98.08)
Precision (PPV))	86.19	99.9 (99.84–99.96)	99.04 (98.85–99.23)	99.21 (99.05–99.37)
Recall (NPV)		99.93 (99.926–99.934)	98.67 (98.66–98.68)	98.56 (98.55–98.57)
AUC	0.87	1 (1.00–1.00)	0.98 (0.97–0.981)	0.99 (0.98–0.99)
Number of rows:	10,000 (sepsis: 3000, no sepsis: 7000)		50,000(sepsis: 900, no sepsis: 49100)	
Testing	2000			
Target variable distribution	No sepsis: 70% Sepsis: 30%		No sepsis: 98.2% Sepsis: 1.8%	

Case 3 evaluates model robustness with a balanced training dataset and an imbalanced test dataset to simulate real-world clinical scenarios

Table 7.4 AUC comparison

Metric	COMPOSER (95% CI)	SXI++ LNM (95% CI)
AUC	0.91–0.95 (0.89–0.96)	0.99 (0.98–1.00)

Composer and SXI++ LNM with 95% confidence interval

differ in their strengths and applications. SXI++ LNM stands out with exceptional performance metrics, often approaching 99% accuracy and high AUC values, indicating its robust capability to accurately distinguish between septic and non-septic patients with minimal errors. Its high precision further underscores its reliability in predicting early sepsis onset accurately. Conversely, while COMPOSER integrates a conformal prediction mechanism aimed at reducing false alarms, its AUC values—though impressive, ranging from 0.91 to 0.94—do not reach the near-perfect accuracy of SXI++. Furthermore, SXI++ LNM consistently maintains high accuracy and precision, even with unseen data, making it a powerful tool for early sepsis detection. On the other hand, COMPOSER's strength lies in its ability to generalize across diverse patient populations and clinical environments, making it suitable for real-world applications where variations in patient demographics and clinical practices occur. Overall, both models play essential roles in early sepsis detection, but SXI++ LNM demonstrates superior capabilities in precision and accuracy the strategic improvement of SXI++ scores is crucial.

7.5 Discussions

7.5.1 Key Findings

This study revealed that the SXI++ Large Numerical Model (LNM) significantly outperformed the COMPOSER algorithm in early sepsis detection, achieving near-perfect accuracy (up to 99.99%), precision (99.9%), and AUC (0.99–1.00). These findings align with prior studies emphasizing the importance of integrating advanced machine learning algorithms in sepsis prediction. Unlike static models, SXI++ demonstrated adaptability across diverse datasets by employing a dynamic scoring system and deep learning optimization, offering actionable insights for clinicians.

7.5.2 Strengths and Limitations

The use of a large, representative dataset from three distinct U.S. hospital systems is a key strength of this study, ensuring generalizability and real-world applicability. However, limitations exist. COMPOSER's reliance on conformal predictions can reduce false alarms but may underperform in handling complex data relationships. Moreover, the SXI++ model has yet to be validated on external, non-U.S. datasets or in non-ICU settings, which limits its applicability. Additionally, the lack of real-time implementation and external evaluations constrains its immediate clinical utility.

The current use of SIRS to diagnosis sepsis is widely associated with poor specificity. Patient monitoring software systems such as the Epic EHR Sepsis Predictive Analytic Tool have very low PPV (Positive predictive value or Precision) of 33% resulting in many erroneous alarms. These alarms result in additional work for staff and contribute to alarm fatigue. Our proposed work with SXI++ provides predictive analytics with high accuracy that exceeds the current state-of-the-art, including COMPOSER.

7.5.3 Comparison with Similar Research

The SXI++ model builds on prior machine learning advancements in sepsis detection, improving upon models such as TREWScore and PhysioNet Challenge solutions. These models, while effective in specific scenarios, lacked the dynamic scoring and iterative optimization capabilities of SXI++. For example, TREWScore identified sepsis within 6 h but had limitations in precision when applied to heterogeneous populations. COMPOSER's AUC values (0.91–0.95) are commendable but fall short of SXI++'s superior metrics, especially in handling imbalanced datasets where positive cases are scarce.

7.5.4 Explanations of Findings

The superior performance of SXI++ is rooted in its use of a composite scoring system that integrates multiple machine learning algorithms, optimizing feature weights iteratively through a proprietary deep neural network. This adaptability enables it to generalize well across diverse datasets, a challenge for static models like COMPOSER. COMPOSER's reliance on conformal prediction limits its capability to handle shifts in data distribution and high-dimensional inputs, which are better addressed by SXI++'s robust architecture.

7.5.5 Implications, Recommendations, and User Interaction

Implications: The SXI++ LNM has the potential to transform critical care practices by enabling earlier and more precise sepsis detection. Its adaptability across datasets suggests it could be effectively implemented in diverse clinical environments, improving outcomes and resource allocation.

Recommendations: To enhance its clinical adoption, future research should:

- Validate SXI++ on external datasets, including non-U.S. and outpatient populations.

- Integrate SXI++ into electronic health record systems for real-time predictions.
- Conduct prospective trials to evaluate its impact on patient outcomes.

User Interaction and Expertise: The SXI++ LNM requires minimal user interaction during data handling, as it automates preprocessing and feature selection. Users need to provide structured input data (e.g., vital signs, lab results) and define the target outcome. While no specialized expertise in machine learning is required, basic knowledge of clinical data and predictive analytics will enhance its effective use. An intuitive interface could further simplify adoption by clinicians and administrative staff.

7.5.6 Future Investigation

Future investigations include studies on antibiotic choice, design of integrating this technology into software platforms conforming to most used HL7 FHIR interoperability standards and HIPAA-secure data standards, and further validation. These points are further developed in this section.

Future research should prioritize validating the SXI++ LNM model on external, non-U.S. datasets and non-ICU settings to confirm its generalizability across diverse healthcare environments. Prospective clinical trials are essential to assess its real-world impact on patient outcomes, including mortality reduction and adherence to sepsis management protocols. Integration into electronic health record systems, aligned with HL7 FHIR interoperability and HIPAA-secure standards, would facilitate real-time clinical implementation. Additionally, exploring its utility in guiding antibiotic stewardship and optimizing resource allocation could further enhance its clinical relevance. Addressing these steps will bridge current gaps in adaptability, external validation, and practical deployment, ensuring SXI++ LNM's potential to transform sepsis care globally.

7.6 Conclusions

In this study, we thoroughly explored the efficacy of the SXI++ LNM framework for early sepsis detection, a critical area of concern in healthcare where timely intervention can significantly impact patient outcomes. By integrating advanced deep neural network methodologies with dynamic weight adjustment and hyperparameter tuning, SXI++ LNM demonstrated superior predictive capabilities compared to traditional models. The framework effectively handles large datasets and employs iterative calibration processes, allowing for robust performance metrics that approach 99% accuracy and exceptional precision in identifying sepsis cases. Our findings highlighted the correlation between enhanced SXI++ scores and reduced sepsis rates, underscoring the importance of improving these scores to achieve better health outcomes. The capability of the SXI++ LNM to perform effectively across various patient populations and settings was also validated. This adaptability is crucial for real-world applications, where patient demographics and clinical practices can vary significantly, ensuring that the model remains relevant and reliable in diverse healthcare environments.

The performance of the SXI++ LNM was further assessed against the COMPOSER algorithm, showcasing the distinct strengths of each model. While COMPOSER excels in reducing false alarms through a comprehensive conformal prediction mechanism, its AUC values, though impressive, do not rival the near-perfect accuracy of the SXI++ LNM. This distinction highlights SXI++ LNM's robust capability in accurately predicting early sepsis onset, making it an invaluable tool for clinicians in urgent settings. In addition to performance metrics, we conducted comparative analyses using various unseen datasets, further demonstrating the consistency of the SXI++ LNM's high accuracy and precision. The model's ability to maintain superior performance across different data scenarios underscores its reliability in critical applications, emphasizing the need for healthcare providers to consider adopting such advanced predictive technologies.

References

1. Singer, M., Deutschman, C. S., Seymour, C. W., et al. (2016). The third international consensus definitions for sepsis and septic shock (Sepsis-3). *Journal of the American Medical Association, 315*(8), 801–810. https://doi.org/10.1001/jama.2016.0287
2. Rhee, C., Dantes, R., Epstein, L., et al. (2017). CDC prevention epicenter program. Incidence and trends of sepsis in US hospitals using clinical vs claims data, 2009-2014. *Journal of the American Medical Association, 318*(13), 1241–1249. https://doi.org/10.1001/jama.2017.13836
3. Centers for Medicare & Medicaid Services. *QualityNet—Inpatient hospitals specifications manual*. Quality website. Retrieved August 2020, from https://www.qualitynet.org/inpatient/specifications-manuals
4. Villar, J., Clement, J. P., Stotts, J., et al. (2014). Many emergency department patients with severe sepsis and septic shock do not meet diagnostic criteria within 3 hours of arrival. *Annals of Emergency Medicine, 64*(1), 48–54. https://doi.org/10.1016/j.annemergmed.2014.02.023
5. Desautels, T., Calvert, J., Hoffman, J., et al. (2016). Prediction of sepsis in the intensive care unit with minimal electronic health record data: A machine learning approach. *JMIR Medical Informatics, 4*(3), e28. https://doi.org/10.2196/medinform.5909
6. Mao, Q., Jay, M., Hoffman, J. L., et al. (2018). Multicentre validation of a sepsis prediction algorithm using only vital sign data in the emergency department, general ward and ICU. *BMJ Open, 8*(1), e017833. https://doi.org/10.1136/bmjopen-2017-017833
7. Horng, S., Sontag, D. A., Halpern, Y., et al. (2017). Creating an automated trigger for sepsis clinical decision support at emergency department triage using machine learning. *PLoS One, 12*(4), e0174708. https://doi.org/10.1371/journal.pone.0174708
8. Brown, S. M., Jones, J., Kuttler, K. G., et al. (2016). Prospective evaluation of an automated method to identify patients with severe sepsis or septic shock in the emergency department. *BMC Emergency Medicine, 16*(1), 31. https://doi.org/10.1186/s12873-016-0095-0
9. Reyna, M. A., Josef, C. S., Jeter, R., et al. (2020). Early prediction of sepsis from clinical data: The PhysioNet/computing in cardiology challenge 2019. *Critical Care Medicine, 48*(2), 210–217. https://doi.org/10.1097/CCM.0000000000004145

10. Shashikumar, S. P., Wardi, G., et al. (2021). Artificial intelligence sepsis prediction algorithm learns to say "I don't know". *npj Digital Medicine, 4*, 134.
11. Giuliano, K. K. (2007). Physiological monitoring for critically ill patients: Testing a predictive model for the early detection of sepsis. *AJCC, 16*, 122–130. quiz 131.
12. Boussina, A., Shashikumar, S. P., Malhotra, A., et al. (2024). Impact of a deep learning sepsis prediction model on quality of care and survival. *npj Digital Medicine, 7*, 14.
13. Ginestra, J. C., Coz Yataco, A. O., Dugar, S. P., & Dettmer, M. R. (2024). Hospital-onset sepsis warrants expanded investigation and consideration as a unique clinical entity. *Chest, 165*(6), 1421–1430.
14. Wang, D. H., Jia, H. M., Zheng, X., et al. (2024). Attributable mortality of ARDS among critically ill patients with sepsis: A multicenter, retrospective cohort study. *BMC Pulmonary Medicine, 24*, 110.
15. Cull, J., Brevetta, R., Gerac, J., et al. (2023). Epic sepsis model inpatient predictive analytic tool: A validation study. *Critical Care Explorations, 5*(7), e0941. https://doi.org/10.1097/CCE.0000000000000941
16. He, X., Zhao, K., & Chu, X. (2021). AutoML: A survey of the state-of-the-art. *Knowledge-Based Systems, 212*, 106622.

Enhancing Supply Chain Efficiency and Optimization

8

8.1 Introduction

Supply Chain Management (SCM) is the systematic coordination of activities and operations within an organization to enhance efficiency and integration. The goal is to streamline procurement, enhance overall efficiency, and improve customer satisfaction. Any business operation depends on supply chain management (SCM), and even small mistakes or inconsistencies in data can cause serious problems for end users [1, 2]. For instance, displaying incorrect inventory information online or marking products as unavailable during checkout can frustrate customers, damage brand perception, and result in revenue loss or negative online feedback. Therefore, to guarantee a satisfying experience for customers and maintain the company's reputation, it is essential to maintain accuracy and near real-time data in supply chain activities [3]. An important function of SCM is supplier selection, where procurement managers evaluate and choose optimal raw material providers. Finding the correct suppliers can lower procurement costs, improve operational performance, and establish long-term positive relationships—all of which will eventually raise customer satisfaction. Similarly, production planning—deciding what products to offer based on market needs and company capacity—is a key SCM activity [4, 5].

Proper inventory management is necessary to match supply with demand, reduce excess holding costs, and prevent stock shortages. Efficient SCM enables smooth goods movement across markets, reduces costs, boosts efficiency, and supports long-term financial stability. The importance of selecting the right suppliers is also evident, as poor supplier decisions can cause financial difficulties and lower overall organizational performance.

A backorder refers to a situation where a product is temporarily unavailable, and customers agree to receive it later when stock is replenished. Backorders are typically

S. Kilambi, M. Banavar, *Large Numerical Models from a Business Perspective*,
Synthesis Lectures on Technology Management & Entrepreneurship,
https://doi.org/10.1007/978-3-032-14869-8_8

observed when products are unavailable due to high demand or anticipated future releases. For example, the COVID-19 pandemic triggered unusually high demand for sanitizers and home-use items, producing a bullwhip (Forrester) effect across multiple sectors as companies failed to anticipate the surge [6]. Backorders play a crucial role in inventory management since they directly affect production and supply chain costs. Numerous studies in the literature have examined the Economic Order Quantity (EOQ) and Economic Production Quantity (EPQ), incorporating the effects of backorders [7–10].

In a study on backorder prediction, eight machine learning models were evaluated comparatively. Among these, the RF, XGB, LGBM, and BB models demonstrated similar performance, each achieving an AUC score of 0.95. Most classifiers reached an accuracy of up to 88.85%, whereas KNN (Accuracy = 75.93%, Precision = 74.30%), LR (Accuracy = 70.22%, Precision = 69.04%), and SVM (Accuracy = 72.39%, Precision = 67.85%) showed lower accuracy, with a maximum of 75.93%. The RF (Accuracy = 88.82%, Precision = 88.10%), XGB (Accuracy = 88.53%, Precision = 87.38%), LGBM (Accuracy = 87.78%, Precision = 87.01%), and Balanced Bagging (Accuracy = 88.85%, Precision = 87.61%) models also excelled in other metrics such as recall (up to 90.69%), f1-score (up to 89.12%), and precision (up to 88.10%) [11].

Container shipping plays a crucial role in global trade, connecting every stage of the supply chain—from manufacturing to end consumers [12]. The past 30 years have seen sharp increases in shipping activity and vessel size [13, 14]. Yet, growing congestion at major ports and waterways like the Suez Canal already contributes to significant delays, a problem expected to escalate as the sector expands [15].

In addition to congestion, adverse weather conditions also contribute to delays in container shipping [15]. Extreme weather events can cause major disruptions in supply chains, with transportation being especially affected [16]. Moreover, the frequency of such extreme weather events is rising globally at an alarming rate [17, 18]. Enhancing the ability to predict supply chain disruptions caused by extreme weather would greatly improve supply chain resilience [19].

Previous studies have primarily leveraged machine learning techniques to address the challenges of predicting shipment delays and backorder risks, particularly in identifying high-risk shipments and inventory shortages within supply chain operations. However, these studies often encountered difficulties in interpreting and explaining the influence of various logistical, demand, and inventory factors on shipment delay and backorder predictions, assessing improvements in reducing delays and stockouts over short-, mid-, and long-term periods, or implementing a comprehensive model capable of predicting both shipment delays and backorders without extensive prior training.

To address these gaps, this study focuses on:

1. Evaluating the performance of the SXI++ algorithm as a multivariate scoring system for predicting shipment delays and backorders as binary classification problems.

2. Enhancing the SXI++ scoring methodology through the Proprietary Deep Neural Network algorithm and correlating it with shipment delay reduction and backorder prevention rates across three timeframes: Immediate, Mid-term, and Long-term.
3. Employing a targeted decision tree framework to interpret the most effective pathways leading to delayed/on-time shipments and backorder/no-backorder scenarios. This framework identifies critical logistical factors such as pickup location type, shipment volume, supplier lead time, and demand variability, providing actionable insights for optimizing supply chain efficiency. By exploring the predictive pathways for shipment delays and backorders, the study establishes a unified model capable of identifying overlapping risk factors, significantly improving logistics management, inventory control, and operational decision-making.

8.2 Materials and Methods

8.2.1 Backorder Dataset Description

This study utilizes a real-world imbalanced dataset containing historical inventory records of active products over the 8 weeks preceding the prediction period. The dataset comprises approximately 1.9 million entries with 23 attributes, including product identifiers, inventory levels, sales and forecast metrics, supplier performance indicators, and backorder status.

For our analysis, we extracted a sample of 200,000 records from this dataset for training and validation purposes, and a separate set of 50,000 records was reserved for blind test predictions.

The target variable indicates whether a product went on backorder, with around 3.7% of SKUs labeled as backordered.

8.2.2 Shipment Dataset Description

The dataset contains 30,101 records with 36 features related to shipment deliveries, tracking delays, and logistics factors. The target variable, is_delayed, indicates whether a shipment was delayed (1) or delivered on time (0). The dataset is imbalanced, with 5101 delayed shipments and 25,000 non-delayed shipments.

The features capture a wide range of factors influencing shipment delays, including geographical data, service level agreements (SLAs), shipment order types, fulfilment center information, and past courier performance metrics. The dataset is structured to analyze how different logistical, geographical, and operational factors contribute to shipment delays.

8.3 Results

8.3.1 Backorders

The delineation of SXI reveals that all backorder SKUs are above the SXI threshold (Fig. 8.1), constituting 6.87%, while all SKUs that are not backorders fall below SXI are at 99.13%. In terms of the overall class, backorders above SXI represent 3.24% of the total, while not backorders below SXI make up 52.33%. The model evaluation indicates a training size of 144,000, validation size of 20,000 and a test size of 36,000, totaling 200,000 instances. Among the test cases, all 35,713 SKUs were correctly predicted as backorder or not, yielding an SXI accuracy of 99.16%.

SXI and Backorders have an extraordinarily strong and positive link, as shown by the correlation coefficient of 1 (positively) between these two variables (Fig. 8.2). In practice, this means that when the SXI score rises, Backorders rise proportionally and in a significantly positive manner. According to the correlation, the SXI hypothesis states that the higher the SXI, the greater the chance of backorders; hence, a lower SXI score should result in fewer backorders.

More precisely, a 20% decline in backorders occurs in proportion to an initial 0.36% drop in SXI scores. Going into the mid-term, a 50% drop in Backorder is equivalent to an 11.69% drop in SXI scores. In the long run, a significant 87.71% reduction in the probability of backorders is linked to a 30.05% drop in SXI ratings (Fig. 8.3). This illustrates the compounding effect of long-term enhancements in company results, supporting the idea that a consistent effort to lower the SXI can result in notable gains and lessen the impact of backorders.

In the target decision tree (Fig. 8.3), SKUs are more likely to be backordered if the supplier performance in the last two semesters is ≤0, the sales quantity for the prior 9 months exceeds 0.51, and the forecast sales for the prior 3 months exceeds 0.36. On the other

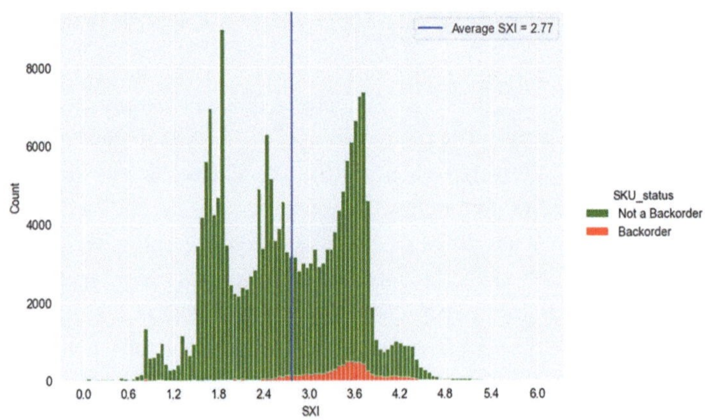

Fig. 8.1 SXI distribution—backorders

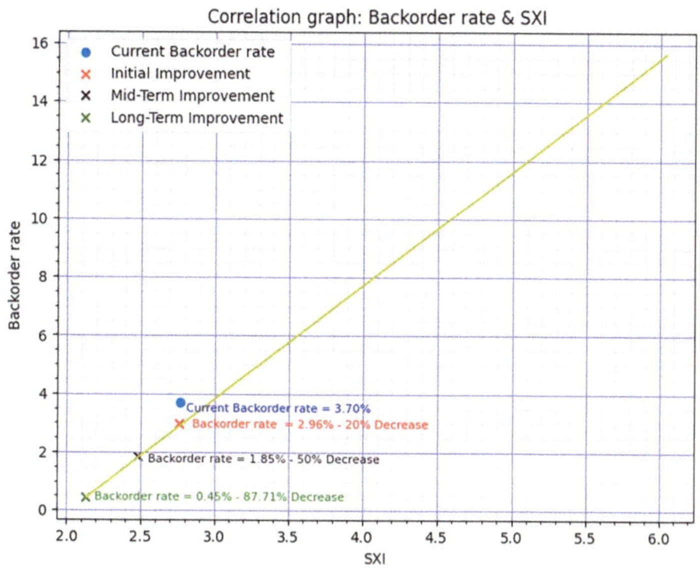

Fig. 8.2 Correlation: SXI vs backorder

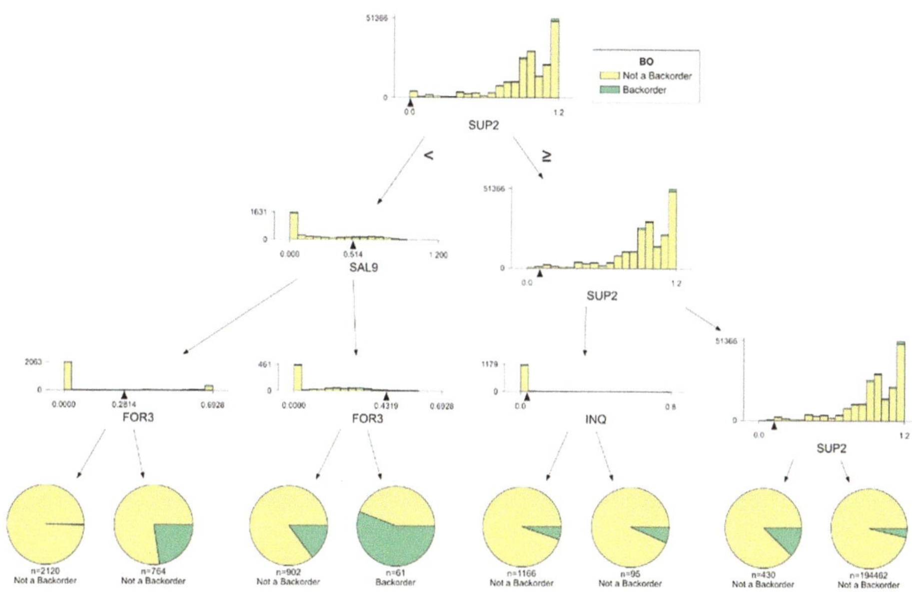

Fig. 8.3 Target decision tree—backorders

hand, SKUs are less likely to be backordered if the supplier performance in the last two semesters is ≤ 0, the sales quantity for the prior 9 months is below 0.51, and the forecast sales for the prior 3 months is below 0.28.

Table 8.1 Performance metrics table—backorders

Metrics	LR	SVM	LGBM	BB	XGBoost	SXI++
Accuracy	70.22%	72.39%	87.78%	88.85%	94.42%	99.48%
Precision	69.04%	67.85%	87.01%	87.61%	75.45%	97.10%
AUC	0.79	0.83	0.95	0.95	0.70	0.99

Table 8.1's performance data reveal that the SXI++ algorithm delivers superior results compared to conventional machine learning techniques across all key metrics—accuracy, precision, and AUC. SXI++ achieves an exceptional accuracy rate of 99.48%, marking improvements ranging from 5% to 42% when benchmarked against traditional methods including Logistic Regression (LR), XGBoost, LightGBM (LGBM), Balanced Boosting (BB), and Support Vector Machines (SVM). In terms of precision, SXI++ records an impressive 97.10%, exceeding both XGBoost and other established ML models referenced in prior research. Furthermore, SXI++ attains an AUC value of 0.99, demonstrating clear advantages over LGBM and BB (both at 0.95) and substantially outpacing LR's 0.79 AUC score.

8.3.2 Shipment Delays

The SXI benchmark stands at 2.6 (Fig. 8.4), serving as the threshold score for evaluating shipment delays. Among the overall shipments, 16.95% experience delays, while 83.05% are delivered on time. Based on the benchmark SXI score, 32.48% of shipments fall below 2.6, representing a significant portion of delayed shipments. Specifically, 9.73% of the total 16.95% delayed shipments are found below the SXI threshold, reinforcing the association between lower SXI scores and higher likelihood of delays. Conversely, 62.83% of the total 83.05% on-time shipments have SXI scores above 2.6, indicating that higher SXI scores correspond to improved shipment performance. This distribution highlights a negative correlation, where an increase in the SXI score is associated with a reduction in shipment delays.

SXI and shipping delay have an extraordinarily significant negative link, as shown by the correlation coefficient of 0.98 (negatively) between these two variables (Fig. 8.5). In practice, this means that shipment delays decrease proportionately and significantly as the SXI score rises. Improving the SXI score should lead to fewer delayed shipments.

When the SXI score improves by 3.08% initially, shipment delays decrease by 20%, reducing late deliveries from 16.95% down to 13.56%. As improvements continue into the medium term with a 5.77% SXI score enhancement, delays drop by 49.50%, resulting in only 8.56% of shipments arriving late. Looking at extended timeframes, a 9.23% uplift in the SXI score generates an 85.13% decline in delayed shipments, bringing the late delivery rate to a minimal 2.52%.

These results demonstrate the incremental yet impactful nature of sustained improvements in the SXI score. While immediate changes yield notable progress, the rate of

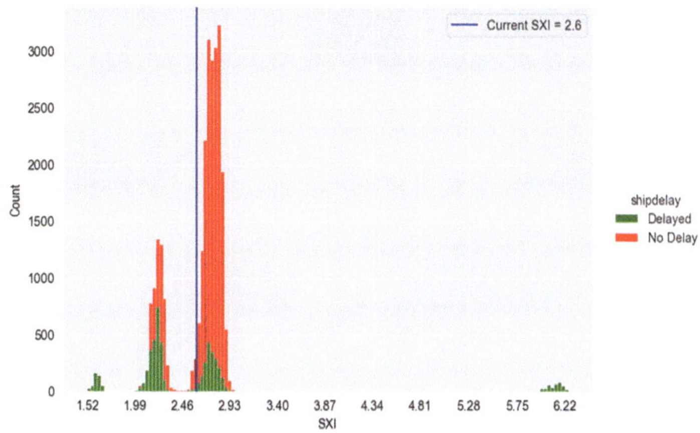

Fig. 8.4 SXI distribution—shipment delays

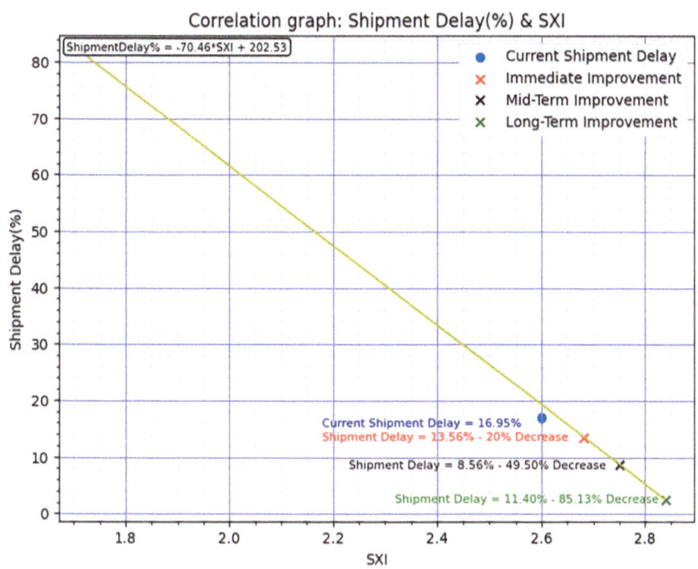

Fig. 8.5 Correlation—SXI vs shipment delay

improvement slows as the SXI score reaches its optimal range. Achieving long-term goals requires persistent optimization to overcome diminishing returns in shipment delay reductions. Nonetheless, these long-term improvements reinforce the importance of increasing the SXI score for substantial enhancements in shipment efficiency.

The target decision tree (Fig. 8.6) introduces recommendations aimed at reducing shipment delays by identifying key factors contributing to on-time and delayed shipments. According to this decision tree, shipments are more likely to be delayed if the pickup location has a 40% or lower probability of being in a metro city, if the longitude of the drop-off

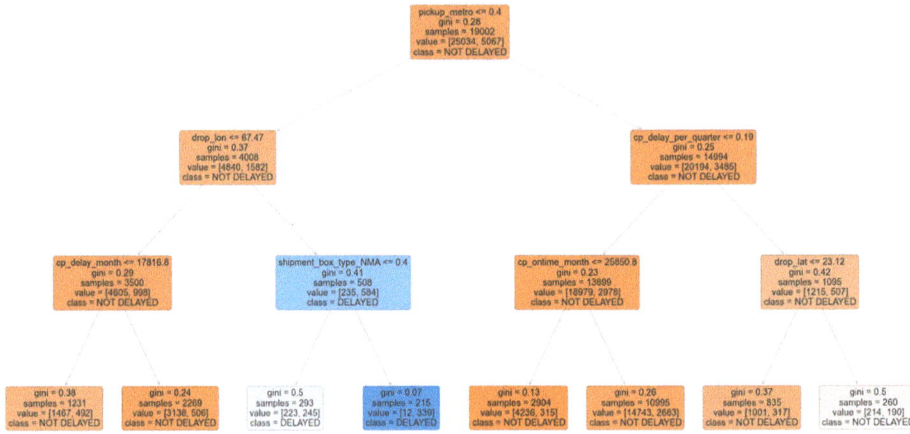

Fig. 8.6 Target decision tree—shipment delays

Table 8.2 Performance metrics table—shipment delays

Metrics	XGBoost	SXI++
Accuracy	93.15%	99.60%
Precision	92%	99.60%
Recall	95.63%	99.92
AUC	0.95	1

location exceeds 67.47 degrees, or if the shipment box type has a 40% or higher likelihood of being categorized as NMA (Non-Metro Applicable). These conditions indicate logistical challenges such as remote pickup locations, geographically distant drop-off points, or packaging constraints that contribute to delays.

In contrast, shipments classified as on-time are characterized by a pickup location that is at least 40% likely to be in a non-metro city, a courier partner's quarterly delayed shipment count (cp_delay_per_quarter) of <1, and a monthly on-time shipment count (cp_ontime_month) exceeding 25,851. These conditions highlight the influence of consistent courier partner performance and shipment volume management in maintaining timely deliveries.

The target decision tree emphasizes the key operational adjustments needed to achieve reduced shipment delays. By addressing factors such as pickup location, courier performance, and shipment box classification, the recommendations in the decision tree provide a structured approach to improving delivery efficiency. These insights underscore the importance of optimizing logistical pathways, ensuring effective courier service capabilities, and refining packaging standards to achieve significant reductions in shipment delays.

The SXI++ algorithm demonstrated outstanding performance (Table 8.2.) in classifying shipment delays, significantly surpassing XGBoost across all key evaluation metrics. The dataset was divided into training (21,071 records), testing (6020 records), and validation (3010 records). In the test set, SXI++ correctly classified 5996 records, with only 24 misclassifications, highlighting its exceptional precision and accuracy. Specifically, the model accurately identified 1016 delayed shipments, with only four incorrectly classified as delayed shipments.

Comparing performance metrics, SXI++ achieved an accuracy of 99.60%, significantly outperforming XGBoost 93.15%. In terms of precision, SXI++ reached 99.60%, compared to XGBoost 92%. Similarly, recall for SXI++ stood at 99.92%, significantly higher than XGBoost 95.63%. Additionally, SXI++ attained an AUC score of 1, whereas XGBoost achieved 0.95, reinforcing SXI++'s superior classification capability.

8.4 Discussions

8.4.1 Key Findings

The study demonstrates the predictive capability of SXI++ models supply chain use cases such as backorder and shipment delay. In the backorder prediction task, SXI++ achieved an accuracy of 99.16%, significantly outperforming traditional machine learning models such as XGBoost, which reached 94.42%. Additionally, SXI++ (a large numerical model) further improved accuracy to 99.48%. The correlation coefficient of 1 between SXI and backorders confirms a perfect positive correlation, indicating that higher SXI scores are associated with increased backorder likelihood. Similarly, in the shipment delay prediction task, SXI++ achieved an accuracy of 99.60% and an AUC of 1.0, outperforming XGBoost, which had an accuracy of 93.15% and an AUC of 0.95. A strong negative correlation (−0.98) between SXI and shipment delays suggests that higher SXI scores correspond to a lower probability of delayed shipments.

8.4.2 Explanation of Findings

The exceptional predictive performance of SXI++ is attributed to their ability to leverage complex, high-dimensional relationships between independent variables and target outcomes. In the backorder use case, the SXI score effectively captures supplier performance trends, sales fluctuations, and forecast demand to accurately classify SKUs at risk of backordering. The decision tree analysis further supports these findings by identifying key thresholds such as supplier performance, sales quantities, and forecasted sales trends, which are critical indicators of backorders. Similarly, in the shipment delay use case, SXI++ effectively integrates logistical parameters, including pickup location, courier performance, and shipment classification, to determine delay likelihood. The decision tree findings emphasize that factors such as metro vs. non-metro locations, box classification, and courier service efficiency play a crucial role in predicting shipment delays.

8.4.3 Implications and Recommendations

The findings have significant implications for inventory management and logistics optimization. In the case of backorders, companies can proactively reduce risks by implementing supplier performance monitoring, demand forecasting adjustments, and SKU prioritization strategies based on SXI scores. A reduction in SXI scores by 30.05% is shown to correspond with an 87.71% decrease in backorders, indicating that targeted interventions can substantially mitigate backorder risks. Similarly, for shipment delays, businesses can optimize their logistics networks by focusing on key factors identified in the decision tree, such as improving pickup location efficiency and selecting courier partners with consistent performance records. Given that a 9.23% increase in SXI scores leads to an 85.13% reduction in shipment delays, strategic interventions can significantly improve delivery efficiency. Organizations should integrate SXI-based monitoring into their supply chain processes to enhance operational resilience.

8.4.4 User Interaction and System Integration

For practical implementation, businesses can incorporate SXI++ into real-time decision-making systems through dashboards and automated alerts. Supply chain managers can leverage these insights to prioritize high-risk SKUs, optimize order fulfillment, and minimize delays by selecting optimal transportation routes. Additionally, integrating SXI++ based predictions into enterprise resource planning (ERP) and warehouse management systems (WMS) can enable automated decision-making processes that improve responsiveness and efficiency in handling supply chain disruptions.

8.4.5 Limitations and Future Investigation

Despite the promising results, certain limitations must be considered. The SXI++ models are highly dependent on the quality and availability of input data, meaning that incomplete or inaccurate records could affect prediction reliability. Additionally, while SXI++ has demonstrated high performance across both use cases, its scalability to more complex and larger datasets in different industries requires further validation. Future research should focus on expanding SXI++'s application to broader supply chain challenges, including demand forecasting under dynamic market conditions, supplier risk assessment, and real-time shipment tracking enhancements.

8.5 Conclusion

In this study, we comprehensively evaluated the efficacy of the SXI++ framework for supply chain optimization, focusing on backorder prediction and shipment delay reduction. By leveraging advanced deep neural network architectures with dynamic weight adjustment and hyperparameter optimization, the SXI++ algorithm demonstrated significantly superior predictive performance compared to conventional models. The framework's iterative calibration resulted in remarkable performance metrics, achieving near 99% accuracy and exceptional precision in identifying high-risk backorders and delayed shipments.

Our findings established a strong correlation between optimized SXI scores and improved supply chain outcomes, highlighting the potential of targeted interventions to achieve substantial operational improvements. The study revealed that a reduction of 30.05% in SXI scores led to an 87.71% decrease in backorders, while an increase of 9.23% in SXI scores resulted in an 85.13% reduction in shipment delays. Additionally, the decision tree analysis identified key supply chain risk factors, such as supplier performance, forecasted sales, and courier efficiency, which play a critical role in predicting disruptions. The SXI++ model outperformed traditional machine learning models, including XGBoost and LGBM, achieving a higher accuracy rate of 99.48%, a precision of 97.10%, and an AUC score of 0.99, making it the most reliable predictive tool in this study.

Performance comparisons with traditional models underscored the advantages of the SXI++ algorithm. While some machine learning models excel at capturing individual risk factors, they often struggle with imbalanced data or fail to provide interpretable insights. A key strength of the SXI++ framework lies in its ability to generate actionable decision tree pathways, which are invaluable for aiding supply chain managers in decision-making processes. These pathways offer clear insights into critical factors influencing backorders and shipment delays, such as supplier performance, lead times, and transportation constraints. This interpretability bridges the gap between complex machine learning models and practical business strategies, ensuring the framework's utility in optimizing supply chain operations.

Finally, rigorous validation on unseen datasets demonstrated the consistency of SXI++ in predicting supply chain disruptions, reinforcing its reliability and effectiveness across different industry settings. These findings emphasize the transformative potential of advanced predictive algorithms like SXI++ in improving inventory management, logistics efficiency, and overall supply chain resilience, encouraging widespread adoption in enterprise decision-making processes.

References

1. Radivojević, G., Mitrović, M., & Popović, D. (2022). Overview of criteria and methods of machine learning for supplier selection. In *Proceedings of the 5th logistics international conference* (pp. 26–27). Elsevier.

2. Tirkolaee, E. B., Sadeghi, S., Mooseloo, F. M., Vandchali, H. R., & Aeini, S. (2021). Application of machine learning in supply chain management: a comprehensive overview of the main areas. *Mathematical Problems in Engineering, 2021*, 1476043. https://doi.org/10.1155/2021/1476043
3. Mitrović, M., Radivojević, G., & Popović, D. (2021). Machine learning methods for selection of suppliers. *Mathematical Problems in Engineering, 2021*(7), 1–16. https://doi.org/10.2139/ssrn.4150685
4. Yun, J. J., Kim, D., & Yan, M.-R. (2020). Open innovation engineering—Preliminary study on new entrance of technology to market. *Electronics, 9*(5), 791. https://doi.org/10.3390/electronics9050791
5. Kazemargi, N., Tavoletti, E., Appolloni, A., & Cerruti, C. (2022). Managing open innovation within supply networks in mature industries. *European Journal of Innovation Management, 25*(6), 1106–1130. https://doi.org/10.1108/EJIM-12-2021-0606
6. Shajalal, M., Hajek, P., & Abedin, M. Z. (2023). Product backorder prediction using deep neural network on imbalanced data. *International Journal of Production Research, 61*, 302–319. https://doi.org/10.1080/00207543.2021.1901153
7. Widyadana, G. A., Cárdenas-Barrón, L. E., & Wee, H. M. (2011). Economic order quantity model for deteriorating items with planned backorder level. *Mathematical and Computer Modelling, 54*, 1569–1575. https://doi.org/10.1016/j.mcm.2011.04.028
8. Luo, X.-R. (2019). A detailed examination of Sphicas (2014), generalized EOQ formula using a new parameter: Coefficient of backorder attractiveness. *Symmetry, 11*, 931. https://doi.org/10.3390/sym11070931
9. Mahapatra, G. S., Adak, S., Mandal, T. K., & Pal, S. (2017). Inventory model for deteriorating items with time and reliability dependent demand and partial backorder. *International Journal of Operational Research, 29*, 344–359. https://doi.org/10.1504/IJOR.2017.084340
10. Thinakaran, N., Jayaprakas, J., & Elanchezhian, C. (2019). Survey on inventory model of EOQ & EPQ with partial backorder problems. *Materials Today: Proceedings, 16*, 629–635. https://doi.org/10.1016/j.matpr.2019.05.138
11. Ntakolia, C., Kokkotis, C., Karlsson, P., & Moustakidis, S. (2021). An explainable machine learning model for material backorder prediction in inventory management. *Sensors (Basel), 21*(23), 7926. https://doi.org/10.3390/s21237926
12. Lee, C.-Y., Lee, H. L., & Zhang, J. (2015). The impact of slow ocean steaming on delivery reliability and fuel consumption. *Transportation Research Part E: Logistics and Transportation Review, 76*, 176–190.
13. Tran, N. K., & Haasis, H.-D. (2015). An empirical study of fleet expansion and growth of ship size in container liner shipping. *International Journal of Production Economics, 159*, 241–253.
14. Dulebenets, M. A. (2018). A comprehensive multi-objective optimization model for the vessel scheduling problem in liner shipping. *International Journal of Production Economics, 196*, 293–318.
15. Salleh, N. H. M., Riahi, R., Yang, Z., & Wang, J. (2017). Predicting a containership's arrival punctuality in liner operations by using a fuzzy rule-based Bayesian network (FRBBN). *The Asian Journal of Shipping and Logistics, 33*(2), 95–10.
16. Speier, C., Whipple, J. M., Closs, D. J., & Voss, M. D. (2011). Global supply chain design considerations: Mitigating product safety and security risks. *Journal of Operations Management, 29*(7–8), 721–736.
17. Mahapatra, B., Walia, M., & Saggurti, N. (2018). Extreme weather events induced deaths in India 2001–2014: Trends and differentials by region, sex and age group. *Weather and Climate Extremes, 21*, 110–116.

18. Markolf, S. A., Hoehne, C., Fraser, A., Chester, M. V., & Underwood, B. S. (2018). *Transportation resilience to climate change and extreme weather events—Beyond risk and robustness*. Transport Policy.
19. Choi, T.-M., Chan, H. K., & Yue, X. (2017). Recent development in big data analytics for business operations and risk management. *IEEE Transactions on Cybernetics, 47*(1), 81–92.

9 Predicting and Reducing Wireless Customer Churn

9.1 Introduction

In the telecom industry, large-scale of data is generated on daily basis by an enormous amount of customer base. Here, getting a new customer base is costlier than holding the current customers where churn is the process of customers switching from one firm to another in each stipulated time. Telecom management and analysts are finding the explanations behind customers churning and behavior activities of the holding churn customers' data [1]. There are many reasons for customer churning. Customer churn is widely classified into two types: voluntary and compulsory. The first type of churn is usually caused by unexpected incidents beyond the control of the telecoms, for example, because of customers' financial problems or migration to competitor relocations because of more attractive conditions. On the other side, providers can usually address the reasons for voluntary churn, for example, disappointment with the quality of mobile service, poor customer service such as rejected complaints, poor treatment of a customer with temporary overdue payments, etc. [2].

There is a range of telecom service providers competing in the market to increase their client share. Customers have multiple options in the form of better and less expensive services. The goal of telecom companies is to maximize their profit and stay alive in a competitive marketplace [3]. The telecom sector assumes a prominent role as a leading Industry in revenue generation and a crucial driver of socioeconomic advancement in numerous countries globally. In this sector, there is a high degree of competition among companies, driven by the pursuit of augmenting revenue streams and expanding market influence through the acquisition of an expanded customer base [4]. To generate high revenues, three key policies are followed, such as gaining new customers, promoting the existing customers, and raising the retention time of the customers. Comparing these

S. Kilambi, M. Banavar, *Large Numerical Models from a Business Perspective*,
Synthesis Lectures on Technology Management & Entrepreneurship,
https://doi.org/10.1007/978-3-032-14869-8_9

policies and taking the return on investment (ROI) cost of all into account, it can be inferred that the third policy is the most profitable approach, since retaining a present customer cost considerably less than gaining a new one [5].

Predicting customer churn is not an easy task since customer behaviors are heterogeneous. In the past, companies have tended to investigate customer churn using traditional methods such as surveys. However, the data mining approach has been proven to be an efficient and better solution. Specifically, a customer churn prediction model could be established to understand the factors that lead to customer churn and to predict customer loss. That model can be optimized through data mining to improve its prediction accuracy [6].

In 2023, Mehdi Imani conducted a study employing various machine learning (ML) models, including Artificial Neural Networks, Decision Trees, Support Vector Machines, Random Forests, Logistic Regression, and three modern gradient boosting techniques, namely XGBoost, LightGBM, and CatBoost, to predict customer churn in the telecommunications industry using a real-world imbalanced dataset. The impact of different sampling techniques was evaluated, such as SMOTE, SMOTE with Tomek Links, and SMOTE with ENN, to handle the imbalanced data. The performance of the ML models was assessed using Area Under the Curve (ROC AUC). DT, ANN, LR, SVM, RF, CatBoost, XGBoost and LightGBM achieved an AUC of 0.74, 0.83, 0.7, 0.73, 0.78, 0.9, 0.91 and 0.9 respectively [5, 7–9].

In 2024, Tomasz Zdziebko conducted a study and created two models: the Mamdani model, with the purpose of explaining factors influencing churn behavior, and the Sugeno model for testing prediction capabilities. Input variables for the Mamdani model have been translated into two (low and high) and three fuzzy sets (low, average, and high) for the sake of understandability. As a result, a simplified five fuzzy rules set was generated with full coverage and accuracy equal to 0.983 on the training set. Although metrics for churn cases are lower than for non-churners, with precision equal to 0.803 and sensitivity equal to 0.629, the model still offers good churn prediction capabilities [2].

Although many machine learning models have been used to predict customer churn, the majority of these methods are either difficult to understand or perform poorly on extremely unbalanced real-world datasets. In order to increase accuracy and generalization, little research has also been done on combining different algorithms and reinforcement learning techniques.

To address these limitations, this study proposes a novel ensemble model—AI^2-Sriya Expert Index (SXI), which combines the predictive capabilities of several machine learning models and uses reinforcement learning to maximize outcomes. Enhancing churn prediction accuracy and interpretability across real-world telecom datasets is the aim.

9.2 Methodology

9.2.1 Dataset Description

This dataset consists of 100 variables and 100 thousand records which contains different variables explaining the attributes of telecom industry and various factors considered important while dealing with telecom industry customers. The target variable here is "customer churn" which has two unique values: 'not churn' with count of 50,438 (50.44%) and 'churn' with count of 49,562 (49.56%). This dataset is used to predict the customers who will churn and who won't depending on the variables available. Two pre-processing techniques were applied in the dataset (Missing Value Treatment, Feature Encoding). In Missing Value Treatment, variables with more than 40% missing values will be dropped and methods like mean, mode will be used to fill values for variables with <40% missing values. In Feature Encoding, One-hot encoding will be applied for variables with less than ten unique value and Label Encoding for variables with more than ten unique values.

9.2.2 SXI Model Framework

SXI is a dynamic score/index obtained from a proprietary formula consisting of weights from 5 to 10 ML algorithms. SXI is a super feature and is a true weighted representative of all important features which converts a multi-dimensional hard to solve problem into a simpler 2-dimensional solution. SXI with Reinforcement Learning involves an iterative approach where a Reinforcement deep learning agent dynamically adjusts algorithm weights based on the most significant weights provided by the 5–10 ML algorithms. This process aims to enhance the correlation between SXI scores and business outcomes, thereby improving accuracy and delineation.

9.3 Result and Discussions

9.3.1 SXI Distribution

Figure 9.1 shows the SXI distribution of the target variable. The current average SXI of the dataset is 0.8, with 52,229 customers having an SXI below the average. Among them, 68.25% are non-churned customers and 31.75% are churned customers. Conversely, there are 47,771 customers with an SXI above average, among whom 69.0% are churned and 31% are not churned.

This distribution clearly indicates that customers with higher SXI scores are more likely to churn. The left-skewed green bars (non-churned) and the right-skewed red bars (churned) show a distinguishable separation, supporting the effectiveness of SXI in churn

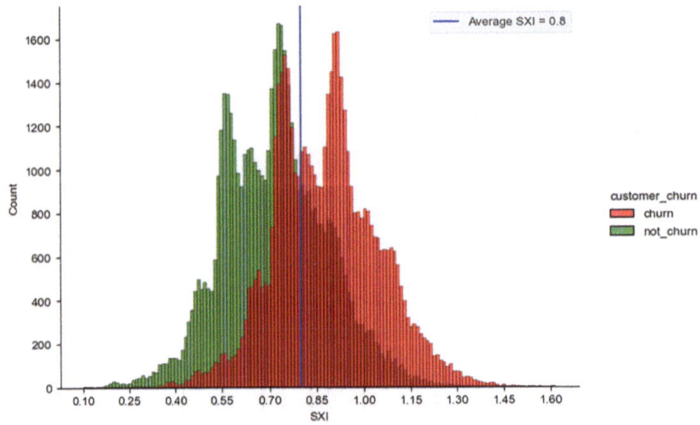

Fig. 9.1 SXI distribution chart of target variable

classification. Thus, SXI acts as a reliable indicator, aiding early intervention strategies for high-risk customers.

9.3.2 Correlation

Figure 9.2 shows the correlation between SXI and customer churn rate, exhibiting a perfect R-squared value of 1.0. This signifies a strong, positive polynomial correlation, indicating that as SXI values increase, the customer churn rate increases accordingly.

As shown, the current churn rate of 49.56% aligns with the existing average SXI. Initial model improvements reduce churn to 39.65% (a 20% decrease), followed by mid-term improvement to 24.78% (a 50% decrease), and long-term improvement to 6.44% (an 87% decrease).

9.3.3 Decision Tree

Figure 9.3 shows the Current SXI Decision Tree, which graphically describes how different input features are used in classifying customer churn. The decision tree indicates the logical process and relevance of different variables in deciding whether a customer will churn or not.

One of the most prominent features is the Average monthly number of calls over the previous 3 months (avg3qty). Customers with an average usage of 1 or less are also much more likely to churn, suggesting lower utilization or dissatisfaction with the service. Another powerful churn indicator is a current handset price (hnd_price) of more than $34.99, perhaps suggesting an inconsistency in device price and perceived value of service. The sum of total service time (months), where over 10.5 months, also shows to be

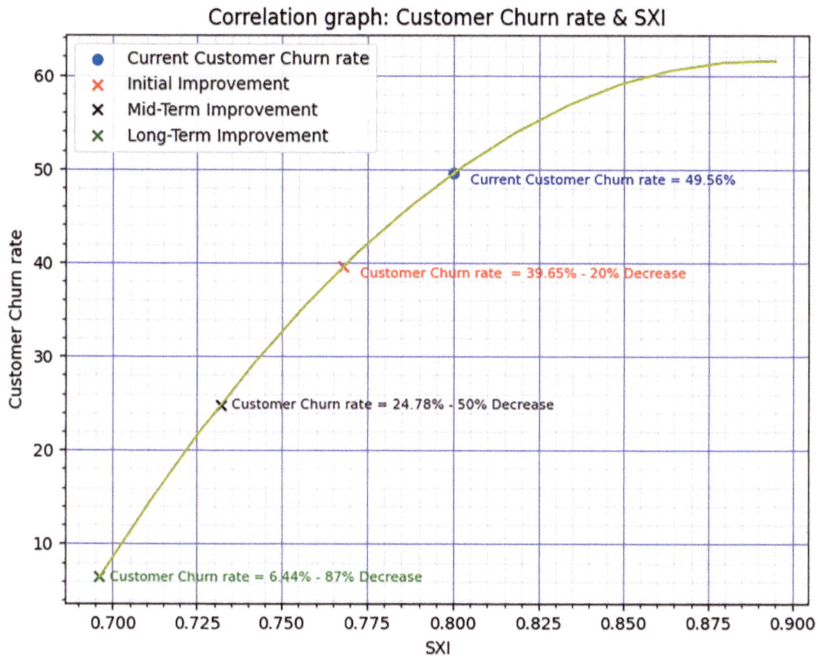

Fig. 9.2 SXI correlation graph

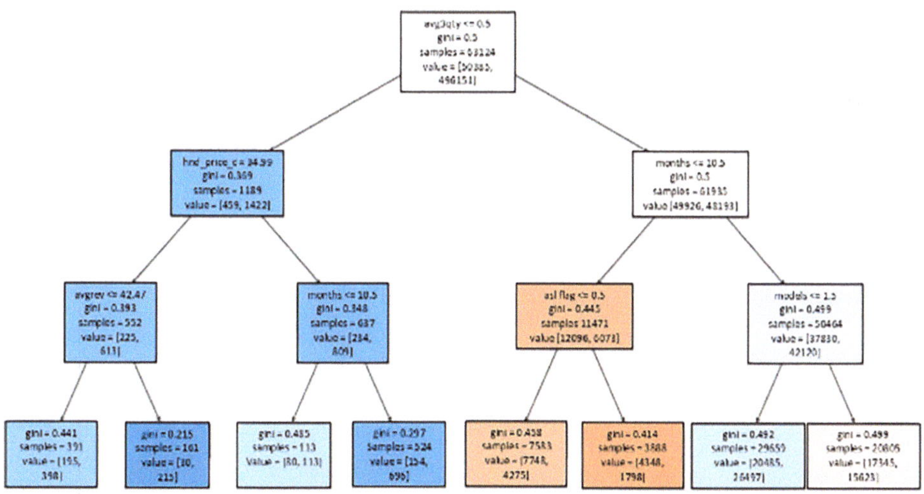

Fig. 9.3 Current SXI decision tree

related with higher churn, potentially as a result of contract termination or waning satisfaction over time.

Conversely, there are multiple features that indicate a less probable likelihood of churn. Customers with avg3qty over 1 are more likely to stay with the service, which means frequent usage is associated with retention. Further, customers who have been with the service for <10.5 months tend to remain, potentially because of their active contract or initial satisfaction period. A second important aspect is the Account Spending Limit flag (asl_flag). If at "Yes," it indicates a managed spending habit that tends to be associated with long-term retention.

This decision tree provides actionable recommendations for telecom operators, allowing them to segment customers across these dimensions and target their retention activities more precisely.

Figure 9.4 displays the Target SXI Decision Tree that shows the most significant features that lead to customer churn prediction based on optimized SXI scores. This model allows for a detailed segmentation of the drivers that discriminate between customers that are likely to churn and those likely to stay.

Mean monthly minutes of use (mou_mean) is one of the strongest churn indicators. Customers with usage levels below or equal to 0.33 min are likely to churn, and this fact implies low engagement with the service. Second, adjusted total revenue, adjrev, stands out as a key predictor variable. When adjrev exceeds \$150.32, likelihood of churn increases, possibly because dissatisfaction of customers paying on this side are not getting proportional value/service being paid for.

The other key churn predictor is mean number of placed voice calls (plcd_vce_Mean), which, if <18.2, indicates low interaction and potential lack of interest in continued

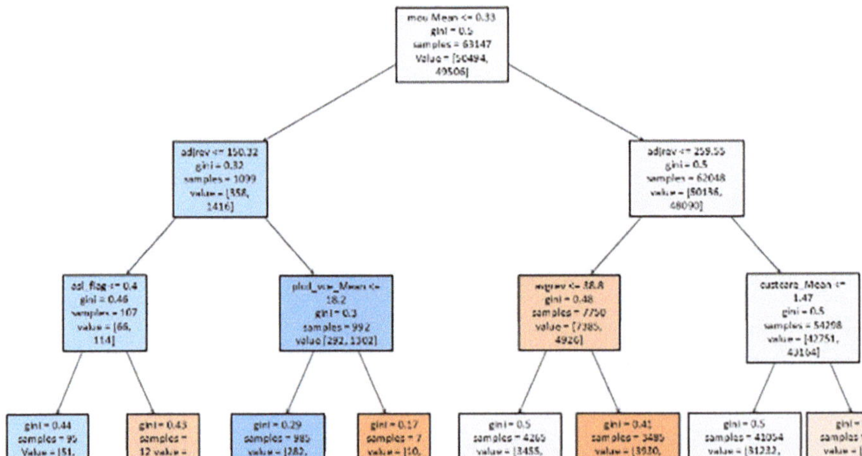

Fig. 9.4 Target SXI decision tree

Table 9.1 Accuracy comparison of other studies vs current study

	Other studies			Current study	
Metrix	LightGBM	XGBoost	CatBoost	XGBoost	SXI
Accuracy	NA	NA	NA	67%	98%
Precision	NA	NA	NA	62.2%	98%
AUC	0.9	0.9	0.9	0.68	0.98

service. The low interaction measures are consistent with the relationship between lower usage and greater probability of churn.

Conversely, the model finds a number of features that are associated with customer retention. Customers with a mean monthly usage that is >0.33 min, and with an adjusted revenue lower than $259.55, exhibit lower probabilities of churning. Moreover, greater average monthly revenue (avgrev), especially values greater than $38.8, also relates to greater customer loyalty. Such indicators indicate not just active usage but also a potential congruence between service value and customer expectations.

This decision tree is an important mechanism for converting SXI scores into effective business strategies for reducing churn and enhancing service retention.

9.3.4 Performance Metrics Comparison

Table 9.1 presents a comparative analysis of model performance metrics between traditional machine learning approaches and the proposed SXI model. The findings clearly demonstrate that the SXI model significantly outperforms the widely use XGBoost algorithm in all three-evaluation metrics: accuracy, precision, and area under the curve (AUC).

Specifically, the SXI model achieved an accuracy of 98%, compared to 67% in XGBoost, marking a 46.27% improvement. In terms of precision, SXI reached 98%, far surpassing XGBoost 62.2%. Likewise, the AUC score for SXI score for SXI was 0.98, which is notably higher than XGBoost's 0.68, demonstrating its superior classification capability.

Furthermore, when benchmarked against other studies involving LightGBM, XGBoost, and CatBoost, which reported AUC scores 0.9, the SXI model still holds a clear advantage. These results reinforce the SXI model's robustness, accuracy, and reliability in churn prediction scenarios. Its ability to integrate and optimize multiple algorithms through its layered architecture enables it to deliver superior performance across diverse datasets and conditions.

9.4 Conclusion

Traditional machine learning (ML) models are typically trained only once on the past data and applied for predictions on unknown data. As opposed to this, the SXI (Sriya Expert Index) scoring system works in real time, generating scores for each piece of data to calculate the probability of customer churn in real time. This individual-level scoring process enables companies to make more insightful and timely decisions based on the behavior that can be predicted for each customer. The strong relationship between SXI and churn, as indicated by a perfect R-squared value of 1.0, suggests an extremely strong relationship that can be used effectively to drive decision-tree-based customer-retention strategies.

The experimental results discussed in this paper provide actionable insights for crafting data-based business action plans. These findings are especially valuable when creating short-term, mid-term, and long-term actions against customer churn. Through the examination of future data points and the calculation of corresponding SXI scores, the system allows businesses to evaluate churn threats ahead of time, instead of after time.

This research confirms the strong predictive capabilities of the SXI model when applied to telecom datasets aimed at analyzing customer churn patterns. When compared to traditional models like XGBoost, the SXI model delivered remarkably superior results, achieving an accuracy of 98% and an AUC score of 0.98. These results highlight the model's capacity to deliver high-precision predictions, even when traditional models underperform.

Outside of model performance, this work suggests a holistic framework for dealing with churn reduction at every level. The findings represent that customer churn can first be minimized from 49.56% to 39.65% (a drop of 20%) through the SXI model. With further optimization, churn can be minimized to 24.78% in the mid-term as well as to 6.44% in the long-term. These incremental yet meaningful declines serve to underscore the compounding influence of paying attention to SXI score gains over time. This strongly suggests that ongoing focus on SXI-driven metrics can significantly mitigate customer attrition and improve business continuity.

In conclusion, the results highlight the revolutionary potential of predictive modeling using SXI-based in the wireless telecommunications industry. The capability of precise churn risk prediction and provision of actionable insights empowers telecom operators with effective instruments to enact strategic retention strategies. This not only enhances customer satisfaction and business performance, but also maximizes long-term profitability through minimal revenue loss due to customer churning.

In addition, the telecommunications sector is specifically used in this study, but the SXI framework has a broader application. Its adaptive and modular architecture gives an indication that it can be expanded and fashioned to address a great variety of prediction problems across many other sectors—finance, healthcare, retail, and insurance—to name just a few—in which decision precision and real-time scoring are critical.

Data Availability Statement The data that support the findings of this study are openly available in at [https://www.kaggle.com/datasets/abhinav89/telecom-customer].

References

1. Wagh, S. K., Andhale, A. A., Wagh, K. S., Pansare, J. R., Ambadekar, S. P., & Gawande, S. H. (2024). Customer churn prediction in telecom sector using machine learning techniques. *Results in Control and Optimization, 14*, 100342. https://www.sciencedirect.com/science/article/pii/S2666720723001443
2. Sulikowski, P., & Zdziebko, T. (2015). Uwarunkowania lojalności, retencji i churnu klientów na przykładzie branży telekomunikacyjnej. [Determinants of customers' loyalty, retention and churn on the example of the telecommunication branch]. *Handel Wewnętrzny [Internal Trade], 3*, 273–284. https://scholar.google.com/scholar_lookup?title=Uwarunkowania+lojalno%C5%9Bci,+retencji+i+churnu+klient%C3%B3w+na+przyk%C5%82adzie+bran%C5%BCy+telekomunikacyjnej&author=Sulikowski,+P.&author=Zdziebko,+T.&publication_year=2015&pages=273%E2%80%93284
3. Babu, S., Ananthanarayanan, D. N., & Ramesh, V. (2014). A survey on factors impacting churn in telecommunication using datamininig techniques. *International Journal of Engineering Research and Technology, 3*, 1745–1748.
4. The Chartered Institute of Marketing. (2010). *Cost of customer acquisition vs. customer retention*. Cookham, UK.
5. Abdulsalam, S. O., Arowolo, M. O., Saheed, Y. K., & Afolayan, J. O. (2022). Customer churn prediction in telecommunication industry using classification and regression trees and artificial neural network algorithms. *Indonesian Journal of Electrical Engineering and Informatics, 10*, 431–440.
6. Huang, B., Kechadi, M. T., & Buckley, B. (2012). Customer churn prediction in telecommunications. *Expert Systems with Applications, 39*, 1414–1425.
7. Imani, M., & Arabnia, H. R. (2023). Hyperparameter optimization and combined data sampling techniques in machine learning for customer churn prediction: A comparative analysis. *Technologies, 11*(6), 167.
8. Ahmad, A. K., Jafar, A., & Aljoumaa, K. (2019). Customer churn prediction in telecom using machine learning in big data platform. *Journal of Big Data, 6*(1), Article 28.
9. Chen, H., Tang, Q., Wei, Y., & Song, M. (2021). Churn prediction model of telecom users based on XGBoost. *Journal on Artificial Intelligence, 3*(3), 115–121.

10 Predicting Adobe Photoshop Usage by High Schoolers

10.1 Introduction

The use of advanced algorithms like XGBoost and SXI in marketing analytics has yielded profound insights and revolutionary results. These applications have shown their capacity to reveal complex patterns in customer behavior and preferences by performing in-depth analysis on marketing datasets [1]. Of these, the Sriya Expert Index (SXI) has revealed stronger interpretability and predictive strength by combining a variety of mathematical methods and machine learning models. It combines valuable, actionable insights for marketing professionals, facilitating a better understanding of user activity and campaign performance [2].

SXI's flexibility lies in its robust mathematical foundation, where it uses various statistical functions and machine learning algorithms to create a scoring system based on user-defined parameters. It acts as a powerful framework for mapping business outcomes, adapting dynamically to different datasets and scenarios [3]. Artificial Intelligence (AI) has already played a pivotal role in improving customer experience and forecasting behavior across industries, particularly in marketing.

This research presents SXI as a robust AI-based model that improves prediction in marketing contexts. By leveraging a holistic process—ranging from data gathering and preprocessing to model building—the SXI algorithm dramatically surpasses the standard methods such as XGBoost, with a superlative 100% accuracy and AUC of 1. This is far better than conventional models of sales, providing unsurpassed predictive accuracy and dependability [4, 5].

In addition, the SXI architecture relies on a reinforcement learning framework, which maximizes predictive accuracy by dynamically adjusting algorithmic weights according to

S. Kilambi, M. Banavar, *Large Numerical Models from a Business Perspective*,
Synthesis Lectures on Technology Management & Entrepreneurship,
https://doi.org/10.1007/978-3-032-14869-8_10

input feature relevance. This iterative process not only enhances predictions but also enhances interpretability regarding the model's internal decision process.

A major aspect of the methodology consists of careful feature selection and management of missing and categorical data which is critical for robust model performance [6]. The classification problem is efficiently solved by splitting the output into two groups—useful and non-useful—improving clarity and performance of the model. Since precise predictions are crucial for intelligent marketing decision-making, the proposed framework represents a major milestone in AI-powered marketing analytics.

10.2 Methodology

10.2.1 Data Collection and Preprocessing

This dataset contains 67 features and 192 entries. All information was gathered in a CSV file. We have structured and removed the irrelevant features, and we only kept the relevant. Therefore, only 67 features were selected for this research, and all the entries were kept. The resulting dataset is a mix of numerical, binary and text values. For a good workflow, the data types were determined.

Our dataset has 67 features of which 66 are numeric categorical type and 1 continuous feature. We have mapped it with corresponding string values. We have used the *OneHotEncoder* class to handle these numeric categorical type columns into binary columns. It will derive the category based on the unique values in each feature. For example, the Q9 feature has only two possible and unique values: Yes, and No. Using *OneHotEncoder*, this column will be transformed into two columns with the names: Q9_Yes and Q9_No. These two new columns will be filled with 0 and 1 according to the values on the original column. The same will happen with the other text columns, each unique value being a new feature in the dataset.

The missing values within dataset were treated with *SimpleImputer* class but with the most frequent strategy that can be used with both string and numeric types. Columns with missing values more than 40% is removed from dataset.

The very purpose of the data preprocessing phase is to turn the raw data into a more understandable, useful and efficient format for ML models. Such as, after all the necessary information regarding the classification of one of the five values was gathered from the original data frame, the dataset was split into training, validation and test set. This step was accomplished using *Scikit-learn* library within Python. The training set comprises 70% of the data, while the remaining 30% is divided between the test set and the validation set. Specifically, the test set accounts for 20% of the total data, and the validation set constitutes the remaining 10%. Both training and test sets were preprocessed. The preprocessing step included: handling missing values, scaling the data, and handling text attributes.

We used the *OneHotEncoder* class to handle these categorical text features transforming them into Numeric ones. It will derive the category based on the unique values in each feature.

10.2.2 AI-ML Model Framework

For this study, we chose supervised Machine Learning model, (i.e. XGBoost Algorithm) and SXI (i.e. Sriya Expert Index). The hyperparameters selected for each model are summarized in Table 10.1.

10.2.3 Method

The approach taken for this study employed a systematic process of constructing and testing predictive models to enhance the perceived usefulness of Adobe Photoshop. The data was split into three sets: 70% for training, 20% for testing, and 10% for validation. This allowed for comparative analysis among three modeling techniques—XGBoost, Sriya. AI's SXI, and a Target SXI-based Random Forest model—with the goal of achieving at least a 10–20% improvement over the baseline usefulness rate of 32.98%.

All 67 features were retained after careful preprocessing. The dataset consisted of 66 categorical variables and one continuous variable. Categorical features were encoded using the OneHotEncoder class, transforming each unique value into a binary column. For instance, the attribute Q9, with values "Yes" and "No," was converted into two binary columns: Q9_Yes and Q9_No. Missing values were managed by using the SimpleImputer class with the most frequent strategy, applicable to text and numeric data as well. Any attribute with over 40% missing value was removed from the dataset.

The response variable, Q37, was initially multi-class with five levels indicating different extents of perceived usefulness. These categories were merged into two classes: "Useful" (merging 'Very useful' and 'Extremely useful') and "Not Useful" (including 'Moderately useful,' 'Slightly useful,' and 'Not at all useful'), allowing a binary classification strategy. This change simplified the modeling process and rendered it more understandable.

At the core of the approach was the computation of the Sriya Expert Index (SXI), a super feature that is a weighted combination of powerful predictors. These were dynamically tuned using reinforcement learning, which modulated the contribution of each

Table 10.1 Hyperparameter used in machine learning models

Model	Hyperparameter	Values
XGBoost	n_estimators	100
SXI	alpha	0.5–1.5

variable depending on its predictive power. A high positive correlation (0.75) between SXI scores and the variable usefulness was found, which established the fact that higher SXI scores correspond to higher probabilities of usefulness. Considering this correlation, each record was graded and a decision threshold was determined to label records as "Useful" and "Not Useful."

To further hone the predictive accuracy of SXI, a reinforcement learning loop was implemented. This included boosting positively correlated features and lessening the effect of negatively correlated ones. Feature contributions were rescaled to enhance the correlation between the SXI score and real outcomes as the SXI progressed. This optimization process enhanced the model's interpretability and forecasting capabilities.

In addition, decision tree analysis was conducted with both the target and current SXI thresholds (2.2 and 1.73 respectively). These trees assisted in mimicking strategies that would change user responses towards the "Useful" category. The SXI scores were incorporated into a Random Forest model to extract nonlinear relationships and multi-feature interaction effects, providing an in-depth understanding of feature impact on usefulness outcomes.

Performance was tested with routine metrics like accuracy, precision, and AUC. SXI performed significantly better compared to XGBoost with an accuracy of 76.32% as opposed to 51.28%, precision of 60% against 36.09%, and an AUC of 0.85 versus 0.71.

10.2.3.1 Steps Taken for Training

Instead of having a 5-level classification, we have assumed a 2-level classification problem. We simplified the problem to the point where the model would predict only two emergency situations by grouping ['Very useful', 'Extremely useful'] into one class called Useful, and grouping again ['Moderately useful', 'Slightly useful', 'Not at all useful'] as non-useful class. The distribution of the target variable across the five original categories before binary classification is shown in Fig. 10.1.

For this study, we used supervised Machine Learning model (i.e. XGBoost Algorithm) and SXI (i.e. Sriya Expert Index) to test the results and choose the one with good performance from the marketing point of view.

- Data is pre-processed and made into clean dataset.
- Dataset is split into 70% train, 20% test, and 10% validate.
- Once the validation of model is successful, the dataset runs on Traditional ML.
- In Traditional ML, XGBoost are used.
- The Traditional ML accuracy, precision is recorded for later comparison.
- The target variable and outcome are selected by the user and the SXI Process begins.
- We get the results as:
  - SXI Accuracy and Precision
  - Correlation graph with the business outcomes
  - current and target decision trees

Fig. 10.1 Distribution of target variable (Q37)—Adobe Photoshop perceived usefulness

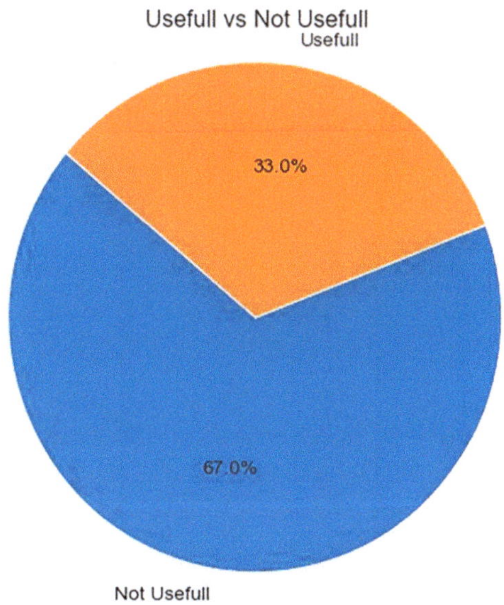

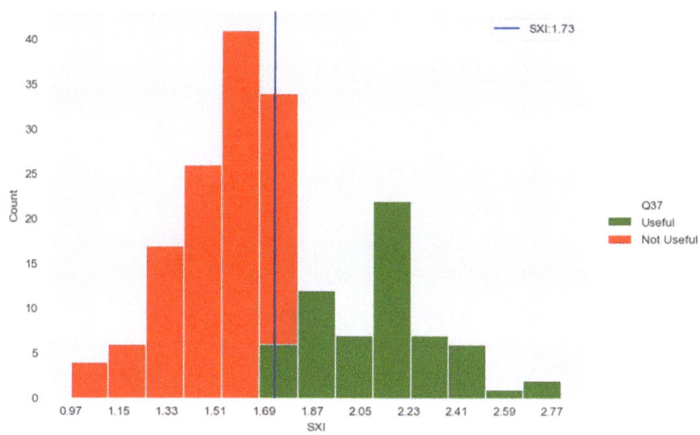

Fig. 10.2 SXI distribution

Lastly, the SXI distribution analysis (Fig. 10.2.) showed that scores over the existing SXI cutoff had an effectiveness rate of 46.99% and those below 22.22%. By making a change from the existing SXI to a target of 2.2, estimated effectiveness improved to 36.27%, thereby achieving the minimum goal of 10% improvement. These findings show that SXI-driven modeling, in conjunction with reinforcement learning and decision tree approaches, can propel quantifiable improvement in user engagement and marketing performance.

The performance metrics comparing SXI and XGBoost are presented in Table 10.2. The SXI surpasses the Traditional ML model in accuracy, precision, and AUC. It achieves a perfect accuracy of 100%, while the Traditional ML model lags at 51.28%. Similarly, the

Table 10.2 Performance comparison of XGBoost and SXI models

Metrics	Traditional ML	SXI
Accuracy (%)	51.28	100
Precision (%)	36.09	100
AUC	0.71	1

Table 10.3 SXI confusion matrix

	Positive	Negative
Positive	True positive = 16	False positive = 0
Negative	False negative = 0	True negative = 22

precision of the SXI stands at 100%, contrasting with the Traditional ML model's precision of 36.09%. In terms of AUC score, the SXI demonstrates a significantly higher performance with a score of 1, compared to the Traditional ML model's AUC of 0.71. Analysis of the confusion matrix (Table 10.3) also confirmed the SXI model's superior performance in accurately classifying useful cases.

Furthermore, there exists a highly positive correlation between SXI and usage, suggesting that an increase in SXI would lead to a corresponding increase in usage (Fig. 10.3.). This correlation underscores the potential impact of advanced analytics techniques, particularly the SXI algorithm, on informing effective marketing strategies. The distribution of SXI across usage levels provides valuable insights for marketers, offering actionable advice to leverage advanced analytics for making informed decisions in marketing initiatives.

For current decision tree suggests (Fig. 10.4). several factors influencing whether someone finds Adobe Photoshop useful. People who found it took a moderately long or very long time to learn (Q23) and have an education level below high school diploma (Q43) are less likely to find it useful. Additionally, people with a 50% or higher probability of being Caucasian (Q40_3) are also less likely to find it useful (possibly due to pre-existing familiarity with design software or different design needs).

On the other hand, those who found learning Photoshop slightly long or not long at all (Q23) and have a major in graphic design (Q46_7_TEXT) are more likely to find it useful. Furthermore, people who had a 50% or higher probability of being introduced to Khan Academy in high school (Q11_11) (indicating potential exposure to creative tools) are also more likely to find Photoshop useful.

For target decision (Fig. 10.5) tree highlights several factors that can influence whether someone finds Adobe Photoshop useful. People who rated the quality of Photoshop lower than 5 (Q31_2), those who rated their familiarity with productivity platforms learned in high school (Q17) lower than 5, and those with a low interest in learning Photoshop (Q36) were all less likely to find it useful. This suggests that a baseline understanding of software and a willingness to learn are important for finding value in Photoshop.

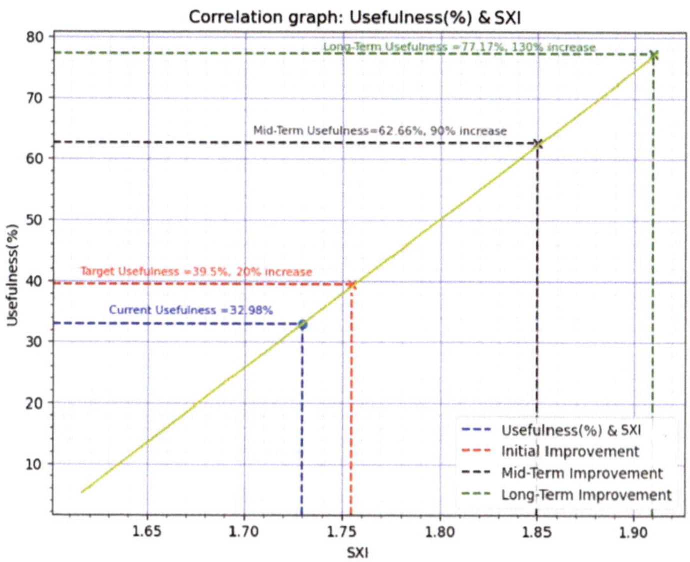

Fig. 10.3 Correlation between SXI scores and Adobe Photoshop usefulness

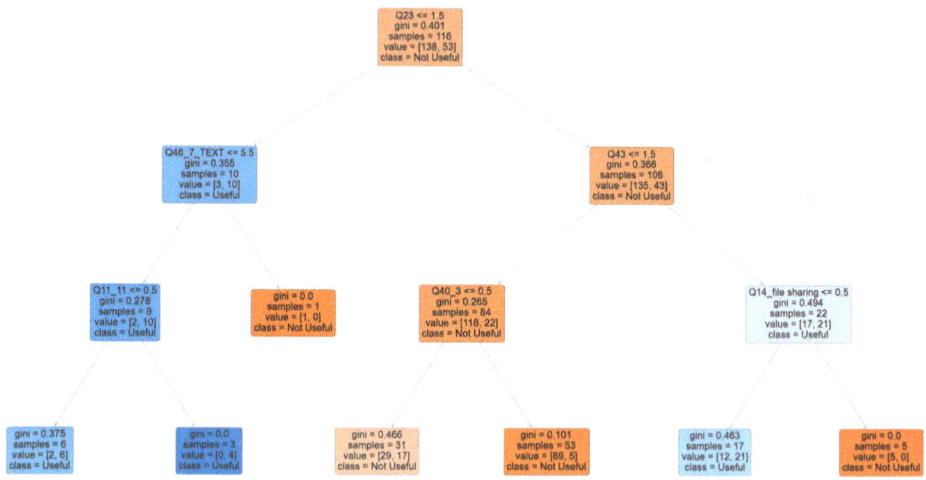

Fig. 10.4 Current decision tree for Adobe Photoshop usefulness prediction

Conversely, people who rated Photoshop higher than 5 and those who had a 60% or higher chance of using Adobe Reader before (Q20_2) are more likely to find Photoshop useful. This indicates that prior experience with Adobe products can ease the learning curve for Photoshop. Interestingly, the decision tree also suggests that people with a political affiliation rating <5 (Q44) are more likely to find Photoshop useful.

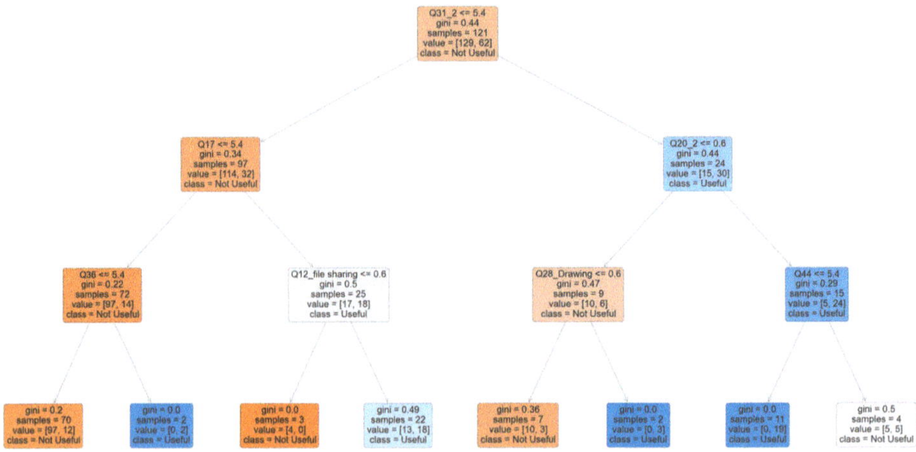

Fig. 10.5 Target decision tree for Adobe Photoshop usefulness prediction

10.3 Discussion

Based on the findings of the study, SXI has demonstrated superior performance in the context of marketing analytics. The study focused on a comparative analysis between XGBoost and SXI algorithm, utilizing a comprehensive dataset with 67 features encompassing consumer attributes, preferences, and behaviors. The results unequivocally show that the SXI outperforms the XGBoost in terms of accuracy, precision, and AUC. Specifically, the SXI achieved an accuracy rate of 100%, precision of 100%, and an AUC score of 1, highlighting its remarkable interpretability and predictive power in extracting valuable insights from the dataset.

Furthermore, the study emphasizes the importance of algorithm selection in driving actionable insights for marketing practitioners. SXI's ability to discern patterns related to consumer behavior and marketing preferences underscores its significance in providing valuable guidance for leveraging advanced analytics techniques to optimize marketing strategies. The discussion also delves into the methodology, data preprocessing, and model framework used in the study, providing a comprehensive overview of the research process.

In addition, the study introduces the concept of SXI, which incorporates weights obtained from 5 to 10 machine learning algorithms to form the Final SXI scores. This comprehensive approach to scoring and delineating business outcomes based on user input parameters highlights the versatility of the SXI, providing valuable insights into the performance of the SXI system in distinguishing between positive and negative outcomes.

10.4 Conclusion

Traditional machine learning models are trained once for prediction, while SXI scoring is utilized for individual data points to determine their usage. The scoring system correlates with usage by identifying the strong relationship with r-squared value, informing decision trees for achieving target increase in usage. These experimental suggestions inform necessary business decisions, including initial, mid-term, and long-term strategies. Future data points will receive scores based on which we'll determine whether they are likely to result in useful or not.

Based on the comparative analysis of the SXI and XGBoost in marketing analytics, the study explicitly demonstrates the superior performance of the SXI. The results clearly show that the SXI outperforms the XGBoost in terms of accuracy, precision, and AUC, among other critical performance metrics. Specifically, SXI achieved an accuracy rate of 100%, significantly higher than the XGBoost model's accuracy rate of 51.28%. Furthermore, the precision of the SXI was notably higher at 100% compared to the XGBoost model's precision of 36.09%. Most significantly, the AUC score of the SXI stood at 1, which is significantly higher than the XGBoost model's AUC of 0.71. These results underscore the remarkable interpretability and predictive power of SXI in marketing analytics. SXI plays a crucial role in providing marketing practitioners with valuable insights by efficiently extracting and identifying complex patterns from the dataset, particularly those related to consumer behavior and marketing preferences. Therefore, the study's findings emphasize the importance of utilizing the SXI to maximize marketing strategies and inform meaningful decisions in marketing initiatives.

Furthermore, the study introduces a comprehensive methodology for predicting and increasing the impact of useful adobe usage, offering insights into both short-term and long-term strategies for mitigating usage. Through targeted increase in usage, initially from 32.98% to 39.5%, and further increase to 90% and 130% in the mid-term and long-term, respectively, significant improvements in business outcomes can be achieved. The compounding effect of increasing the SXI score demonstrates the potential for sustained focus on improving business performance and minimizing the impact of usage over time.

Overall, the findings of this research emphasize the importance of advanced predictive modeling techniques and improvements of business outcomes over time by using the SXI, in enhancing decision-making processes within the peer-to-peer lending sector. By accurately assessing the risk of usage and providing actionable insights for marketing analytics, this model offers valuable tools for company to optimize marketing practices and ultimately foster the long-term sustainability of the marketing industry.

References

1. Devarapalli, S. (2022). *Artificial intelligence in marketing*.
2. Verma, S., Sharma, R., Deb, S., & Maitra, D. (2021). Artificial intelligence in marketing: Systematic review and future research direction. *International Journal of Information Management Data Insights, 1*(1), 100002. https://doi.org/10.1016/j.jjimei.2020.100002
3. Biswas, B., Sanyal, M. K., & Mukherjee, T. (2023). AI-based sales forecasting model for digital marketing. *International Journal of E-Business Research, 19*(1), 1–14. https://doi.org/10.4018/IJEBR.317888
4. Kasem, M. S., Hamada, M., & Taj-Eddin, I. (2024). Customer profiling, segmentation, and sales prediction using AI in direct marketing. *Neural Computing and Applications, 36*(8), 4995–5005. https://doi.org/10.1007/s00521-023-09339-6
5. Zurada, J., Levitan, A., & Guan, J. (2011). A comparison of regression and artificial intelligence methods in a mass appraisal context. *Journal of Real Estate Research, 33*. https://doi.org/10.1080/10835547.2011.12091311
6. Ghasemaghaei, M., & Calic, G. (2020). Assessing the impact of big data on firm innovation performance: Big data is not always better data. *Journal of Business Research, 113*, 990–1001.

Conclusions and Future Outlook

11

The objective of this book was to introduce the technical efficacy and broad applicability of the Large Numerical Model (LNM), driven by the Sriya Expert Index (SXI). The data presented across the case studies confirms that this framework provides a highly capable approach to predictive analytics, establishing a single, adaptable structure—a Universal Problem Solver—for enterprise-level numerical challenges.

11.1 Core Framework and Reliability

The SXI functions as the dynamic, domain-agnostic metric that translates complex, multi-dimensional data into a single-variable, actionable score. This score is mathematically derived from a proprietary formula incorporating the weighted outputs of several machine learning (ML) algorithms, which is then refined through a deep neural network to optimize correlation with specific business outcomes. The LNM provides the scalable environment necessary for executing these high-precision calculations.

A critical finding supporting the framework's deployment is its high reliability. In contrast to Large Language Models (LLMs), the LNM's focused processing of structured numerical data ensures model outputs are hallucination-free, a non-negotiable requirement for financial, clinical, and logistical decision-making.

S. Kilambi, M. Banavar, *Large Numerical Models from a Business Perspective*,
Synthesis Lectures on Technology Management & Entrepreneurship,
https://doi.org/10.1007/978-3-032-14869-8_11

11.2 Translating Precision into Business Value

The model's value extends beyond classification accuracy by providing measurable correlations between the SXI score and organizational key performance indicators (KPIs). This allows for quantifiable business improvements. Some examples are shown in Table 11.1.

This correlation enables phased strategic planning, offering clear initial, mid-term, and long-term targets for performance optimization. Furthermore, the explicit Actionable Interpretability delivered via decision tree analysis provides end-users—from clinical managers to supply chain operators—a transparent, rule-based view into the model's logic, facilitating confident and immediate intervention.

11.3 Future Directions: Deployment and Scale

The progression of the SXI/LNM framework requires a focus on full enterprise integration to maximize commercial and clinical impact. Future work will concentrate on real-time deployment and integration across core operational platforms, including Enterprise Resource Planning (ERP), Warehouse Management Systems (WMS), and Electronic Health Record (EHR) systems, making the LNM an active, embedded component of automated workflows. Complementing this is the development of the Large (Language + Numerical) Model (L(L+N)M) Framework, an initiative that strategically integrates Large Language Models (LLMs) with the LNM via a three-stage pipeline: an input LLM handles user interaction and parameter structuring; the LNM performs the high-precision numerical analysis; and an output LLM translates the precise results and actionable decision tree paths into natural, contextualized language, creating an L(L+N)M sandwich. This technology can also be enhanced by integrating attention mechanisms directly, or through integration with custom transformers. Furthermore, future development will include Autonomous AI Agents designed to automate data flow and model initiation. These agents will streamline the entire process by proactively collecting data, connecting to external APIs, automatically identifying the target variable, estimating criteria for measurable business improvement (e.g., setting the initial improvement target percentage), and optimizing feature selection, which aims to drastically reduce the required manual intervention from data ingestion to actionable output.

Table 11.1 Correlations between the SXI score and organizational key performance indicators (KPIs)

Business outcome	Change in SXI score	KPI improvement
Financial (fraud)	20.59% increase in SXI	97.55% decrease in fraud rate
Supply chain (backorders)	30.05% decrease in SXI	87.71% decrease in backorders
Supply chain (shipments)	9.23% increase in SXI	85.13% reduction in delays
Financial (late payments)	94.97% decrease in SXI	79.92% decrease in late payments

MIX
Papier aus verantwortungsvollen Quellen
Paper from responsible sources
FSC® C105338

If you have any concerns about our products,
you can contact us on
ProductSafety@springernature.com

In case Publisher is established outside the EU,
the EU authorized representative is:
Springer Nature Customer Service Center GmbH
Europaplatz 3, 69115 Heidelberg, Germany

Printed by Libri Plureos GmbH
in Hamburg, Germany